Naturalists' Handbooks 26

Insects on dock plants

DAVID T. SALT
Division of Biological Sciences,
Institute of Environmental and Biological Sciences,
Lancaster University

and

JOHN B. WHITTAKER
Division of Biological Sciences,
Institute of Environmental and Biological Sciences,
Lancaster University

with illustrations by Michael J. Roberts

Published for the Company of Biologists Ltd by

The Richmond Publishing Co. Ltd

P.O. Box 963, Slough SL2 3RS, England

Series editors
S. A. Corbet and R. H. L. Disney

Published by The Richmond Publishing Co. Ltd,
P.O. Box 963, Slough, SL2 3RS
Telephone: 01753 643104

ISBN 0 85546 309 0 Paper
ISBN 0 85546 310 4 Hardcovers

Printed in Great Britain

Contents

Editors' preface

An ecologist who focuses on the animals associated with a common plant can be sure to find interesting animals to study and can begin to unravel the interactions that link together the species in a representative insect/plant community. Many students have had their first experience of field research exploring the fauna of nettles, thistles or cabbages and oilseed rape using books in this series. Dock plants are widely accessible and pleasant to work with – they do not have stinging hairs, prickles or a smell. Their fauna is now well enough known to be accessible to the non-specialist, but it still holds enough unsolved mysteries to challenge the investigator. We hope this book will stimulate further developments in the study of insect/plant communities in general, and dock plant communities in particular, at all levels.

S.A.C.
R.H.L.D.
August 1997

Acknowledgements

Without the help of the following people at various stages of its development this book would not have been produced. Peter Flint has been involved with much of the primary field work and identification of insects for many projects leading up to this publication and has given useful advice, particularly on the keys. Geoffrey Halliday and Andrew Malloch kindly advised on the key to dock species. Keith Harris provided invaluable help tracking down gall midges and Kevin Tuck of the Natural History Museum helped with tortricid identifications. Stephen Judd of Liverpool Museum kindly lent hoverfly specimens. We are grateful to Sally Corbet and Henry Disney for their support during the project and setting exacting standards, and to staff of the biology field station of Lancaster University for help culturing insect and plant material over many years. Maps were prepared by the Biological Records Centre, Institute of Terrestrial Ecology, Monks Wood.

D.T.S
J.B.W.

1 Introduction

This book deals with docks and sorrels, members of the genus *Rumex*. There are 13 species native to the British Isles (Kent, 1992)*. Many of these have a world-wide distribution and some are considered major weeds. A high proportion of the insect species associated with *Rumex* are specialists in that they are restricted to the genus, or at least to the small family (Polygonaceae) to which the plants belong. This may be because the plants tend to have high levels of calcium oxalate in their tissues which may protect them from many generalist feeders. Insects which can feed on dock are of special interest because of their potential as biological control agents.

Because of their accessibility, even within school and college grounds as well as private gardens, docks are an ideal subject for studies of general entomology and the processes of insect herbivory and ecology and offer many unanswered questions. They, and some of their associated insects, are also easy to culture in greenhouse or laboratory.

In this book we have brought together information on those insect species which have a special association with *Rumex* and also some common generalists which may be found feeding on the plants. We have also included the important insect predators and parasitoids of the herbivores, many of which have themselves a close association with the food plant. In all we include in the identification keys over 60 species of insects. Many of these are common and widespread but their relative abundance will vary from year to year and from place to place. We do not include casual visitors to the plants nor other invertebrates such as mites and slugs, although Key II will help in the recognition of these groups. The majority of the species of insects listed can be found on *Rumex obtusifolius* and *R. crispus*, though some, particularly amongst the weevils and aphids, are specialists on the smaller *Rumex* species.

In the first part of the book we introduce the species of *Rumex* and some common and easily studied insect species, and we consider how the insects and plants interact.

In the second part of the book are keys for the identification of species of *Rumex* and insects which are likely to be found on them. Other species are outside the scope of this book and should be identified using a general work such as Chinery (1986). The keys are followed by outline suggestions of techniques for studying these insects and setting up simple experiments on research areas where our knowledge is incomplete.

It is very much our hope that this book will give the reader the confidence to search for further interactions between docks and insects and to undertake novel investigations. It is still the case in entomology that the gifted

* References cited under authors' names in the text appear in full in References and further reading on p. 49.

amateur can make important contributions to knowledge.

Throughout this book we use two-part (binomial) Latin names for the plants and animals. The first (generic) name is often abbreviated to its initial capital letter. By convention, the second (specific) name is always given a lower case initial letter. At first mention, the (abbreviated) name of the authority who first described the species is added. Latin names are sometimes changed as taxonomic knowledge improves, so in older texts the same species may be found with a different scientific name. The insect names in this book follow the latest editions of the Kloet and Hincks (1964–1978) check lists of British insects, but sometimes an earlier name is appended in brackets, if it is still in common use. Only rarely do insects have unambiguous common names but these are used where appropriate for the more familiar groups such as host plants, moths and butterflies. Plant names follow Stace (1991) but common names are also used for all species.

Immature (larval or pupal) stages of insects are frequently found associated with plants because the larva is usually the stage in the life cycle when most feeding and growth take place. Many insect herbivores feed and pupate within the plant tissue, hidden from sight, and some of the immature stages listed here will be found by dissection of the plant tissue.

2 Docks and sorrels

Species of docks

Docks (genus *Rumex*) belong to the family Polygonaceae which includes a number of common weed species in other genera such as Japanese knotweed *Fallopia japonica* (Houtt.), knotweed *Persicaria maculosa* Gray and knotgrass *Polygonum aviculare* L. The name dock is derived from the Anglo-Saxon *docce* or Greek *daukos* which mean a type of carrot or parsnip and presumably refer to the plants' tap-roots. Sorrel, on the other hand, is from the Old French *sorel* meaning sour of taste (Macleod, 1952).

More than fifty species of dock have been recorded in the British Isles. Most of these are scarce, recorded near places such as dockyards, mills and breweries where their seed was accidentally imported from other countries and plants became established. The main species dealt with in this book are common native species: *Rumex acetosa* L. (sorrel), *Rumex acetosella* L. (sheep's sorrel), *Rumex crispus* L. (curled dock), *Rumex hydrolapathum* Hudson (great water dock), which has a largely central and south-eastern distribution (fig. 1), and *Rumex obtusifolius* L. (broad leaved dock) which is widely distributed around the country (fig. 2). Other native docks e.g. *R. aquaticus* L., *R. longifolius* DC., *R. maritimus* L., *R. palustris* Smith, *R. pulcher* L. and *R. rupestris* Le Gall have a much more restricted distribution. Some species (sorrel *R. acetosa* and French sorrel *R. rupestris*) are used as culinary herbs, whilst consumption of others has even resulted in human fatalities. Their use as a common antidote to the alkaline poison of nettle stings arises from their acidic leaf tissue. For example, *R. obtusifolius* contains oxalic acid.

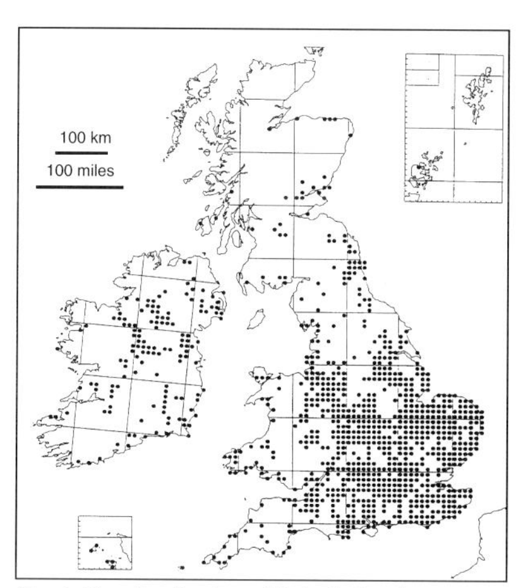

Fig. 1. The distribution of *R. hydrolapathum* in the British Isles.

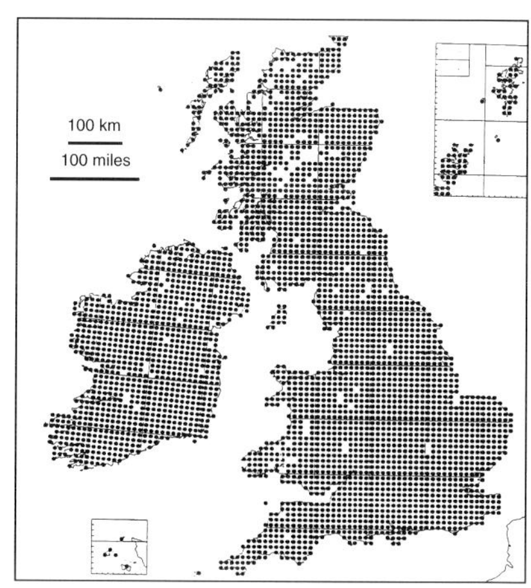

Fig. 2. The distribution of *R. obtusifolius* in the British Isles.

Table 1. *Characteristics of common species of* Rumex.

Scientific name	Common name	Habitat	Status as a weed	Typical height (cm)
Rumex acetosa	sorrel	fertile grassland heavy metal spoil	weeds of meadows contaminant of seed	15–60
R. acetosella	sheep's sorrel	acidic pastures	weed of forestry nurseries and poor acidic pastures	8–25
R. hydrolapathum	great water dock	water edges, often part submerged	unimportant	up to 180
R. obtusifolius	broad-leaved dock	agricultural land and roadsides	important agricultural weed	30–90
R. crispus	curled dock	agricultural land and roadsides	important agricultural weed	30–90
R. conglomeratus	clustered dock, sharp dock	in damp places	unimportant	30–100
R. sanguineus	wood dock	woodland, often near buildings	unimportant	30–100

Many species of docks grow on disturbed ground such as cliff faces, river banks, dunes and waste and arable land. Characteristics of the species which enable them to be successful in these habitats, such as fast growth and prolonged survival of seeds in the soil, also make them tenacious weeds (Salisbury, 1964). The range of habitats they inhabit can vary widely from fertile ground (*R. acetosa*) to infertile ground (*R. acetosella*) and from low to high altitudes. Some species are so important that the Weeds Act, 1959, made it a finable offence to permit the spread of these 'injurious weeds'. It singled out five weed species, the docks *R. obtusifolius* and *R. crispus*, the thistles *Cirsium vulgare* (Savi.) Ten. and *C. arvense* (L.) Scop. and ragwort *Senecio jacobaea* L.

Docks are herbaceous (non-woody) plants and range in height from a few centimetres (*R. acetosella*) to 1–2 metres (*R. hydrolapathum*). The major species are easily distinguished, but where species intermingle, hybrids (which are generally sterile) are often found near the parent plants.

hermaphrodite: having both male and female reproductive organs on the same individual

dioecious: having distinct male and female plants

Docks are generally perennials, that is they survive for a number of years. The larger species such as broad leaved dock and curled dock do not normally flower until the second year of growth. Many are hermaphrodite (for example *R. obtusifolius*) and self fertile and mostly pollinated by wind rather than by insect pollination. Others (such as *R. acetosa*) are dioecious. Species which grow in or near water (such as *R. hydrolapathum* and *R. aquaticus*) can use the water to disperse their seed, whilst in others, like broad leaved dock, the hooked teeth around the fruits attach to mammal coats and disperse their seed. At the end of the growing season the plant dies down often leaving stems, a rosette of green leaves and shrivelled leaves above the soil surface. In the following spring the plant regenerates from the rootstock.

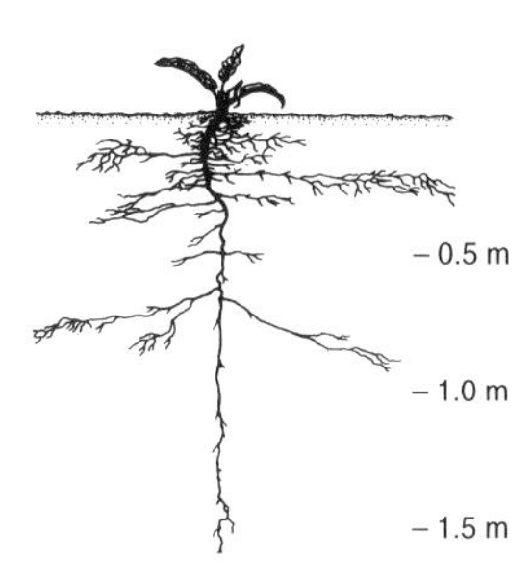

Fig. 3. Root system of *R. crispus* (Salisbury, 1964).

Deep roots of some species (fig. 3) mean that they are resistant to drought. Ploughing and other disturbance can sever roots of many species and these root fragments can then regenerate into new plants. Heavy attack by insects can reduce seed size, leading to fewer viable seedlings. As soil is disturbed, for example by digging or in worm casts, seeds which have been dormant for a number of years are brought to the surface and may germinate. The main periods for germination are in spring and in autumn when the seedlings can grow in the spaces left by other plants, where competition for light and soil nutrients is lowest.

Docks as feeding sites

Insects live in and on plants, as people live in buildings, in many different ways. Thinking of plants as architectural structures with different degrees of complexity has enabled ecologists to understand more about how insects interact with plants, and to make predictions about the communities of insects on the plants.

inflorescence: the flowering shoot produced by a plant

Most docks are built to the same basic pattern, with relatively large leaves, a persistent rootstock and tall open inflorescences. The relative sizes of these organs differ between species and affect the insect species which can utilise them. For example old mines excavated by weevils in stems of *R. obtusifolius* provide hiding places for a range of other invertebrates including earthworms, woodlice and centipedes. Mines in the narrower stems of smaller dock species do not provide as much space as the larger species. Leaf size and overlap also affect the microclimate of the insects on the host plants, and thus different conditions are provided by different species. As docks are a closely related taxonomic group their chemistry is similar. This means that insects which are found on one species will often survive on others. For example, the fauna of *R. obtusifolius* is very similar to that of *R. crispus*.

An insect must consume enough food of the right quality to complete its development on the plant. The smaller species of dock may provide insufficient food for development to be completed. For example, the larvae of leaf-mining flies are at risk from predators when, having exhausted one leaf, they leave it and move to another. In leaves of the smaller species of dock, eggs may be laid singly on the leaves to maximise the chances for survival of the larvae. Every time the larva has to move to another leaf its chances of survival are reduced.

Effects of insects on dock plants

Docks are robust, fast-growing plants which are difficult to eliminate from most habitats in which they occur. It is not surprising, therefore, that they are not usually considered to be particularly susceptible to insect damage. Nevertheless insects can inflict significant damage on *Rumex* species. How much damage can the plants sustain without

instar: the developmental stage of an insect between successive moults

cuticle: superficial layer covering an animal or plant surface

epidermis: outermost layer of cells

root/shoot ratio: the ratio of dry biomass of root to shoot tissue

guard cells: cells that border the pores (stomata) of a leaf and control their function

loss of fitness or ability to contribute to the next generation? The most visible damage to the larger leaved docks (*R. obtusifolius, crispus, sanguineus* and *hydrolapathum*) is caused by chrysomelid beetles (*Gastrophysa viridula* on *R. obtusifolius, crispus* and *sanguineus* and *Galerucella nymphaeae* on *R. hydrolapathum*) because they cause extensive defoliation in their active stages (pl. 6).

First-instar larvae of *G. viridula* remove the cuticle and lower epidermis but older larvae can chew right through the leaf except the largest veins whilst adults, especially gravid females, can remove almost any part of the leaf. The damage resulting from this grazing is not confined to the parts of the plant which suffer it. Heavy damage has the additional consequence of changing the root/shoot ratio (reducing it in *R. crispus*) and reducing both the number and size of seeds produced. Whilst this reduction in seed number is unlikely to affect fitness because a large excess of seeds is produced, seedlings from smaller seeds are less able to compete with other plants (fig. 4) (Cideciyan & Malloch, 1982).

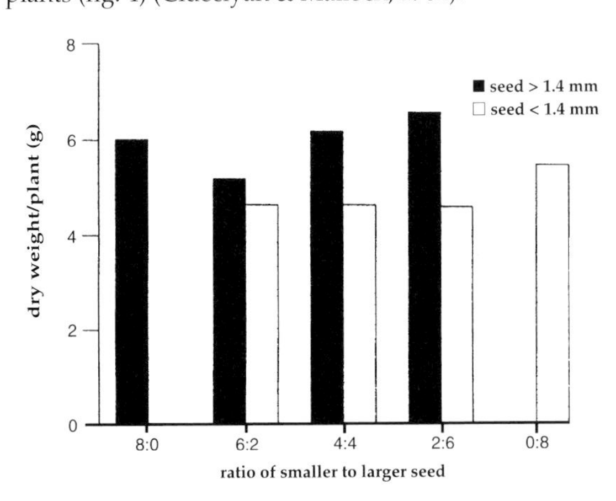

Fig. 4. Effects of competition between *R. obtusifolius* plants grown from different-sized seed on plant weight at harvest (after Cideciyan & Malloch, 1982).

Chewing damage caused by the caterpillars of butterflies and moths (Lepidoptera) and sawflies (Hymenoptera) has a similar effect on the plants. Some of these caterpillars affect root biomass directly by feeding below ground. The next most visible type of damage results from leaf-mining by fly larvae, *Pegomya* species (pl. 4). Indeed, sometimes this can exceed the damage caused by chrysomelids. In a heavy infestation, almost all the leaves on a plant may be mined and each mine can occupy most of the surface area of a leaf. When young and small, the mine is a paler green colour than the normal upper leaf surface. Larvae can be seen moving around within it and leaving trails of black frass (excreta). Eventually the epidermis of the leaf dies and goes brown. During this process the leaf gradually loses its ability to photosynthesise and to regulate water loss because the guard cells which control gas and water exchange die and finally the cuticle itself becomes permeable (Whittaker, 1994).

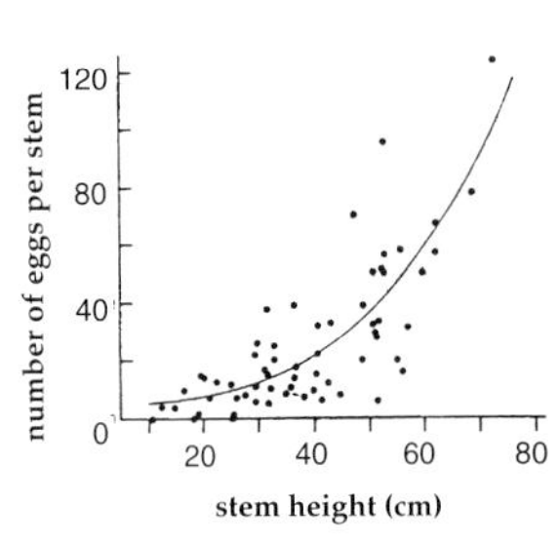

Fig. 5. The relationship between numbers of *A. curtirostre* eggs in stems of *R. acetosa* and stem height (from Hopkins & Whittaker, 1980*a*).

phloem: the system of tubes which transports sugars and amino acids

xylem: the system of tubes which transports water and minerals

parasitoids: insect parasites which usually cause the death of the host

The sap-feeding insects have a less visible effect on the plant, because they remove phloem (aphids) and xylem (froghoppers) sap. This can result in wilting. It may also cause leaf distortion (pl. 6), probably as a consequence of toxic saliva injected by the insects.

Larvae of the weevil species bore inside stems, leaf stalks or roots and the adult weevils and their parasites leave distinctive exit holes (pl. 4, fig. 19). In *Apion curtirostre*, there is a curvilinear relationship between the number of eggs laid per stem by the ovipositing females and the height of *R. acetosa* stems (fig. 5).

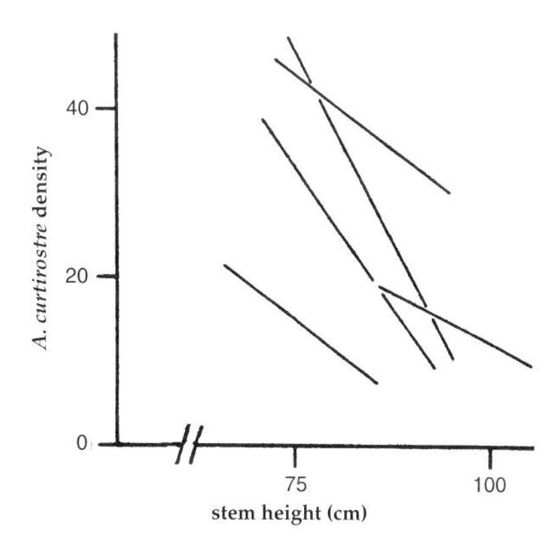

Fig. 6. Effect of *A. curtirostre* density on female stem size of *R. acetosa* (from Hopkins & Whittaker, 1980*a*).

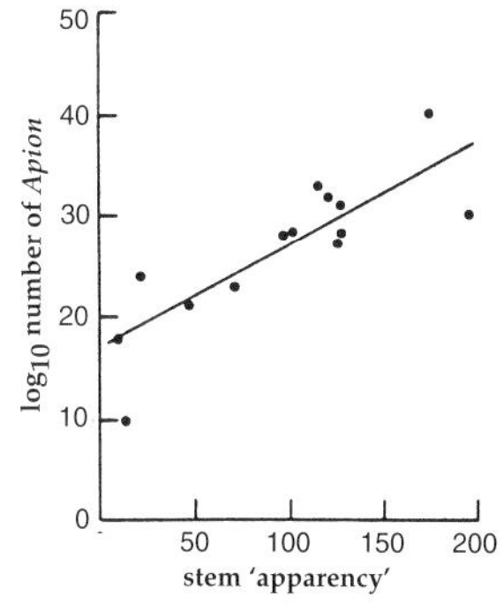

Fig. 7. The relationship between *Apion* numbers and stem 'apparency' (number of stems x stem height x 10) (Hopkins & Whittaker, 1980*b*).

Thick female stems are also preferred to the thinner male ones. In turn, female stem height of *R. acetosa* is reduced as the number of attacking weevils increases (fig. 6). A similar effect has not been observed in male stems.

Rates of attack by the weevil *Apion violaceum* seem to depend on the product of stem numbers and stem height at a particular site (fig. 7). *A. violaceum* feeding sites cover a range of stem diameters in *R. crispus*, but *Apion frumentarium* (= *miniatum*) tends to be found lower down the stem where it is wider (fig. 8). Thicker stems are thought to confer protection from parasitoids (Freese, 1995). In general the highest densities of weevils per stem tend to occur on plots with the most diverse vegetation.

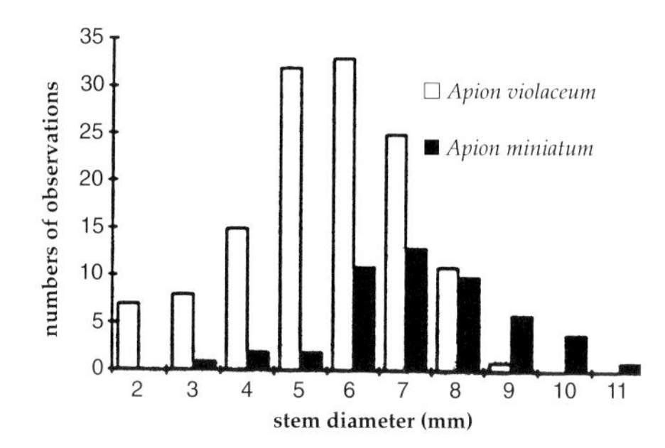

Fig. 8. Stem diameters at the feeding sites of *A. violaceum* and *A. miniatum* in *R. crispus* (Freese, 1995).

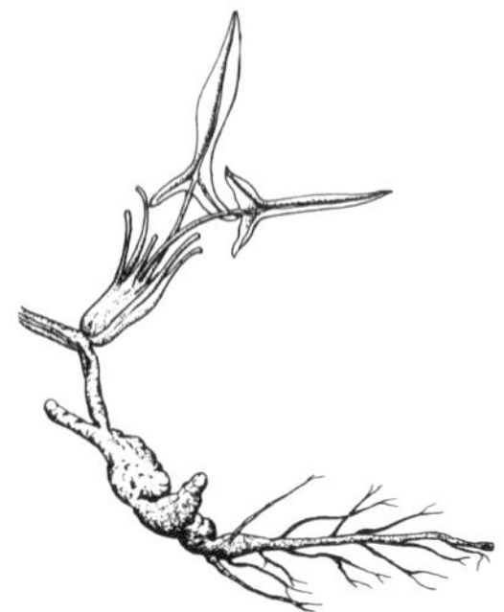

Fig. 9. *Apion sanguineum* gall on the roots of *R. acetosella* (from Docters van Leeuwen, 1982).

Other weevils feeding in the shoots or roots can produce galls in the plant tissue. An example is *Apion sanguineum* in the roots of *R. acetosella* (Docters van Leeuwen, 1982) (fig. 9). Adult weevils often produce small roughly circular holes in the leaves.

The long-term consequences to the plant of insect damage have been most thoroughly studied in the case of *Gastrophysa* feeding. It is rare for insects alone to kill the plant. However, insect herbivory usually acts, not on its own, but in addition to other stresses on the plant. The ability of *Rumex crispus* to compete with *R. obtusifolius* and of *R. obtusifolius* to compete with surrounding grasses are both very much affected if they are simultaneously damaged by insect grazing (Bentley & Whittaker; 1979, Cottam and others, 1986).

Likewise, *R. crispus* on riverside shingle banks in N. Lancashire were found to be unable to survive winter flooding if they had been damaged by *Gastrophysa* grazing the previous summer whereas ungrazed plants survived perfectly well. This was because the beetle grazing had caused the plants to translocate material from the roots to replace grazed leaves, and this had weakened the roots too much, allowing the plants to be uprooted in floods (Whittaker, 1982).

A fungal 'rust' disease *Uromyces rumicis* (Schum.), which produces brown pustules on the leaves, also weakens *Rumex* plants. This infection often occurs at the same time as beetle grazing. Insect herbivory is found to induce resistance of plants to rusting. The rust increases herbivore mortality and decreases fecundity (Hatcher and others, 1994).

Recently, there has been considerable concern about the extent to which elevated concentrations of carbon dioxide in the atmosphere may affect plant growth. What is often ignored is that the delicate balance between plant and insect herbivores may be disturbed. As yet few studies have been made of this, but two (Salt & Whittaker, 1995 and Brooks & Whittaker, in press) are concerned with *Rumex obtusifolius* and its herbivores *Pegomya* and *Gastrophysa*. The effects on the insects are not direct, but are mediated through changes in the plant such as nutrient status and quantities of secondary (defence) plant chemicals like calcium oxalate. Other environmental factors have also been shown to affect the balance. Pollutants such as sulphur dioxide and ozone (fig. 10) or addition of nitrogen either as fertiliser or pollutant often lead to greater damage to the plant, either because it offers a more rich source of food or,

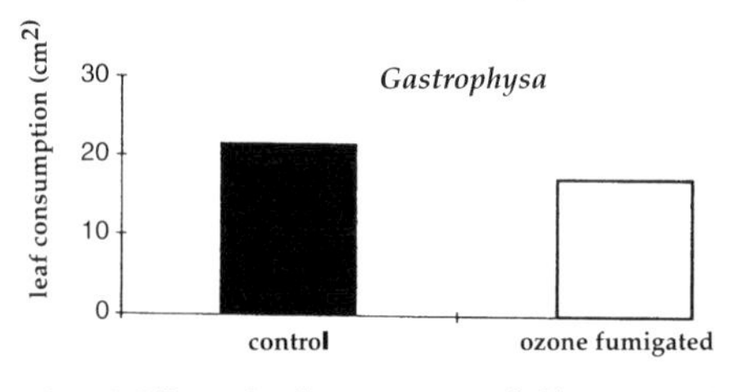

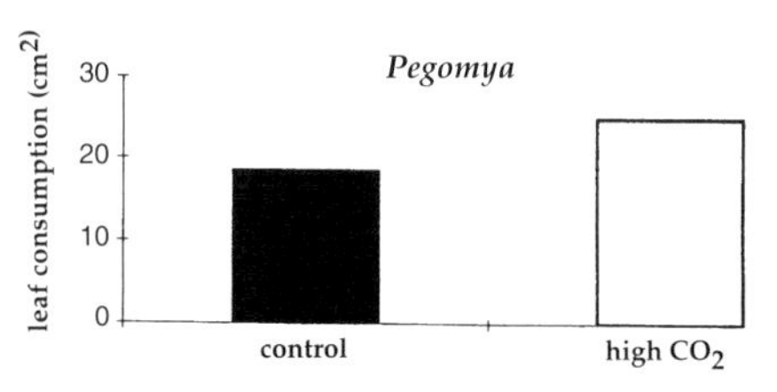

Fig. 10. Effects of pollutant gases on (left) consumption of *R. obtusifolius* by *G. viridula* on ozone-fumigated plants, and (right) mine size of *P. nigritarsis* on plants grown in elevated carbon dioxide.

paradoxically, because it is less nutritious and the insect needs to consume more to survive and reproduce (Whittaker and others, 1989; Salt & Whittaker, 1995).

So interactions between insects and *Rumex* plants and between the plants, the insects and other factors are extremely complex and only just beginning to be unravelled. If these weeds are to be controlled by biological, rather than chemical, means, these are the interactions which will need to be understood. There is a great deal of scope here for further research.

Biological control of dock plants

Two of the large species of docks (*Rumex obtusifolius* and *crispus*) cause problems in agriculture in Europe and North America where they are unpalatable and even toxic to grazing animals and reduce yield of crops. Other species such as *R. acetosa* are common contaminants of the seed of crop plants with which they can spread from site to site. Although herbicides can be used to control these weeds their large rootstocks and the long-lived seeds of some species make them difficult and expensive to control. Seed of some species can remain viable 20 years after being buried in the soil and *R. crispus* seed for 80 years. Large quantities of seed can be produced. A single *R. crispus* plant can produce up to 40,000 seeds per year. Therefore there has been some interest in using the natural herbivores or pathogens of these plants in their control. These natural control agents, which may have partial success, can be combined with herbicides or different cultivation practices to give what is called integrated control. Cutting down the plants often stimulates regrowth and has little effect on seed production. As a result frequently-mown roadside verges often contain large numbers of *Rumex obtusifolius* plants.

Good candidates for biological control (biocontrol) need to cause substantial damage to the weeds so that these are no longer economically important. Also candidate insects need to be specifically dock feeders (Barbattini and others, 1986) which will not transfer to other related crops such as rhubarb (*Rheum rhabarbarum* L.) and in North America buckwheat (*Fagopyrum esculentum* Moench), which was once also cultivated in the British Isles for flour and livestock feed (Foster, 1989; Mabey, 1996). Biocontrol agents need to be able to survive and reproduce easily in the climatic conditions where they are released and be easy to culture in huge numbers in captivity.

Several dock herbivores and pathogens have been suggested which can, in some conditions, result in complete defoliation of the plants and reductions in the weight and number of seeds. These are the beetles *Gastrophysa viridula*, *Hypera rumicis* (De Gregorio and others, 1991) and *Apion frumentarium* (= *miniatum*) and the rust fungus *Uromyces rumicis*. Combinations of rust infection with herbivore damage are thought to stand the best chance for dock control by attacking the plant in different ways. It remains to be seen whether or not dock herbivores can help to halt the spread of Japanese knotweed. More information is needed on the palatability of this plant to dock herbivores, and on the biocontrol potential of insects which might attack different parts of the plant.

Further study

Active, easily visible insects are to be seen on most dock plants from April to October. A number, however, overwinter in some form on, in or near to the plant and can be found with careful searching. Stem borers (mostly weevils) leave signs that can still be found after the insects have matured and abandoned the plant host. Details of where and how to find the more important species are given in the relevant sections on biology. Much can be done by careful observation throughout the spring and summer. It is very instructive to follow a few individual plants through and to record the sequence of insects attacking or utilising the plant in different ways. Fig. 11 shows the times of appearance of some of the species common around Lancaster.

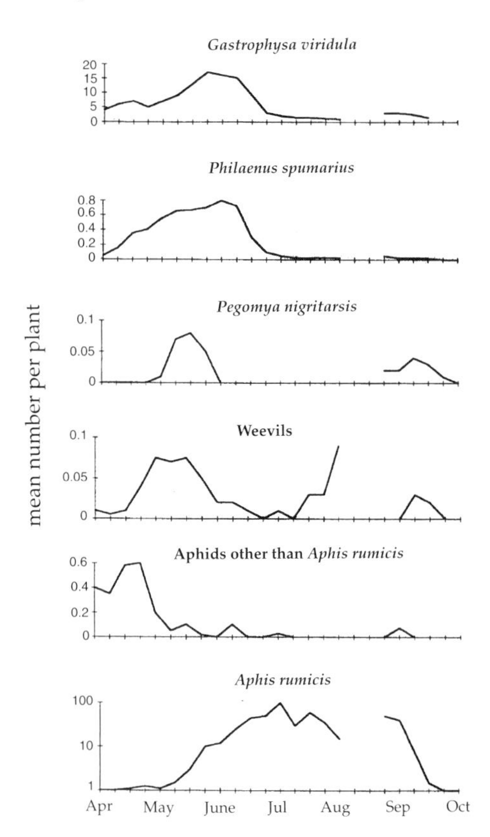

Fig. 11. Sequence of common insect herbivores on *Rumex obtusifolius* at Lancaster.

Many herbivores and predators associated with docks take refuge in dead or curled living dock leaves. Does removal of dead material around the plants increase or decrease the amount of damage to docks? How does it affect the number of insect species living on the plants?

It is known from other insect/plant systems that infestation by one herbivore can deter egg-laying by others (Finch & Jones, 1989). Does infestation by *Gastrophysa* or aphids on docks deter colonisation by dock-feeding flies? Do the ants associated with some aphids on docks deter infestation by other herbivores on the plants? Do plants with ant-tended aphids grow better than plants on which aphids are not ant-tended?

The broad relationships between some of the insects and the plants described in this book are at least partly known. However, the galleries and mines excavated by beetles and fly larvae provide refuges for a range of other insects and invertebrates. Which animals make use of these chambers? How do they interact with other members of the animal community on the plants? How do they affect the host plant?

A quantitative study of the dynamics of the entire community of insects feeding on a *Rumex* species throughout a season would be very valuable. Rarely do we have this kind of information. Such observations might be used to explore the following ecological principles.

(i) Resource partitioning . How do different insect species share out in space and in time the resources offered by dock plants?

(ii) Food webs. Starting with the plant as the primary producer, how many further trophic levels can be supported in different environments? How are these interconnected through food webs? (see, for example, fig. 21).

(iii) Top-down, bottom-up regulation. Are population sizes of herbivorous insects on docks generally regulated by food resources ('bottom-up') or by natural enemies, predators or parasitoids ('top-down')? How does exclusion of the predators and parasites affect plant growth and herbivore performance?

(iv) Species/area relationships. Large clumps of docks seem to support more insect species than small clumps or isolated plants. Which species are only present in the larger patches and which in the smaller?

(v) Defensive chemicals. Do dock plants with higher concentrations of oxalic acid support fewer insects? Oxalic acid can be seen in the crystalline form in leaves by dissolving out the chlorophyll and observing the crystals through a polarising microscope.

3 Biology of key insect species

The spittlebug *Philaenus spumarius* (Homoptera) (pls. 2, 5)

Nymphs

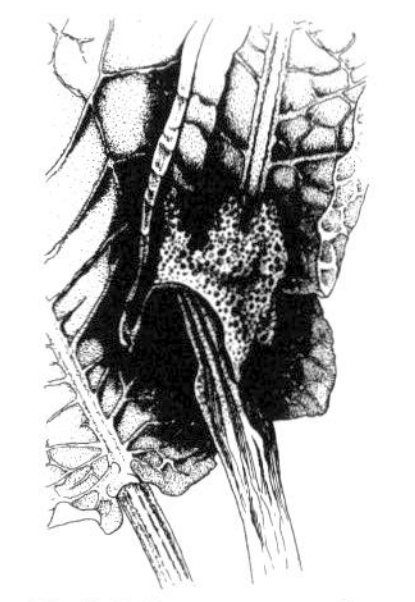

Fig. 12. Spittle mass produced by *P. spumarius* on *R. obtusifolius*.

nymph: immature stage of certain insect groups, such as bugs and grasshoppers, in which there is no pupal stage. (The immature stages of groups such as flies, which do have pupae, are known as larvae)

tracheae: breathing tubes of insects connecting openings (spiracles) to the internal organs of the body

The spittle masses which are to be found on the underside of leaves and on the stems of *Rumex* plants, especially *R. obtusifolius* and *R. crispus*, contain the immature stages (nymphs) of the spittlebug or froghopper, *Philaenus spumarius* (fig. 12). These frothy white masses are produced as a by-product of feeding when the insect inserts the stylet mouthparts into xylem vessels. The nymphs feed head downwards and suck out the xylem sap using a pump situated in the distended face of the insect. Most of the dilute sap is passed out of the gut and drains down into a cavity formed by the overlapping sternal plates of the abdomen. Air is blown into the cavity from the abdominal tracheae to form the bubbles which are fixed by a secretion from glands (often distinctly red) on either side of the abdomen. This species is by no means restricted to *Rumex*. It has been recorded on a very large number of plant species, mostly dicotyledons. The nymphs can be distinguished from those of other members of the group because they lack markings and are yellow to green all over.

The nymphs generally remain within the spittle, so they are easy to observe. The spittle provides protection from desiccation, predators and parasites. There may be more than one nymph in each spittle mass and the cast skins (exuviae) are often found within the spittle. Nymphs in spittle may be found between May and July. Larger spittles indicate older stages towards the end of this period.

Adults

Adult froghoppers emerge from the spittle in June and July and survive until September or October. At this stage they are much more mobile and may leave the host plant for another of the same or a different species. Although they continue to feed on the xylem, spittle is not produced and excess sap is simply ejected from the anus. A major point of interest about this species is that it exhibits about 12 colour forms or morphs (fig. 13) which occur in balanced equilibrium, but with different frequencies of morphs in different habitats. The colour forms are determined in a simple genetic way, but it is still not fully understood what determines their frequencies in different situations. There is some evidence that environmental pollution may play a role since melanic forms appear to occur at higher frequencies in heavily polluted areas. Selection of uncamouflaged

melanic coloration: dark coloration

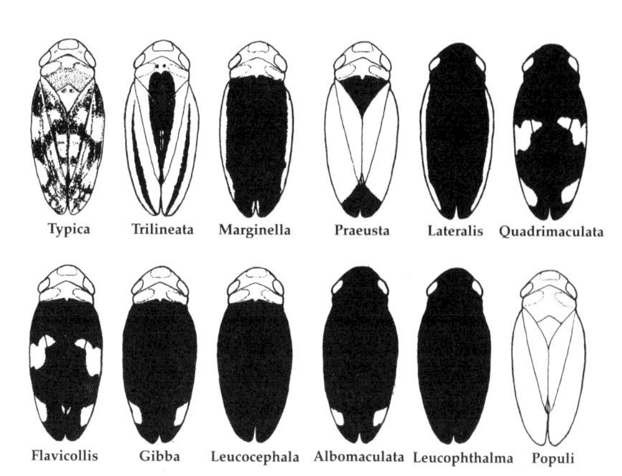

Fig. 13. The colour forms of *Philaenus spumarius* (redrawn from Harper, 1974).

individuals by a visually hunting parasitoid could also be important. This parasite, *Verrallia aucta* (Fallén), is unusual in its group (the fly family Pipunculidae) because it attacks the adult froghopper and not the nymphs, which are presumably too well protected in their spittle (Whittaker, 1969). The parasite (fig. 14) can be found in August and September by dissecting the abdomen of the host, which will eventually be killed when the parasite emerges to pupate in the soil. It is not unusual to find 70 to 80% of froghoppers parasitised.

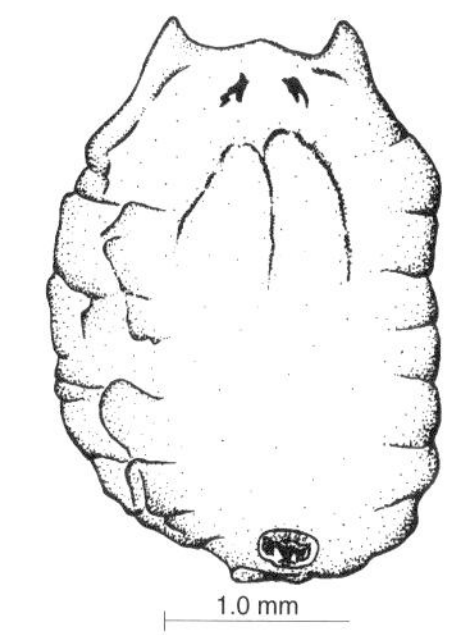

Fig. 14. *Verrallia aucta* larva, parasite of adult *Philaenus spumarius*.

Some unanswered questions

Adult froghoppers may be collected from *Rumex* plants in July to October by sweep net, pooter (fig. 25) or by beating the plant and dislodging the insects onto a cloth or tray. They can be separated into colour morphs (fig. 13) and by sex (fig. 15). If the morphs are classified as melanics (dark forms), forms without markings (populi), and typica and trilineata, frequencies of the commonest morphs can be compared in different situations. Are the colour morphs found on *Rumex* simply a sample of those in the surrounding habitat, or are certain morphs more or less frequent on *Rumex* than elsewhere? Are they selecting *Rumex* to feed on, or surviving better on *Rumex* than on other plants? Is the percentage of melanics higher in areas of industrial pollution than in unpolluted countryside? What proportion of the froghoppers on *Rumex* are parasitised? Is this the same as in the surrounding populations of froghoppers on other host plants? Are all morphs and both sexes equally susceptible to parasitism?

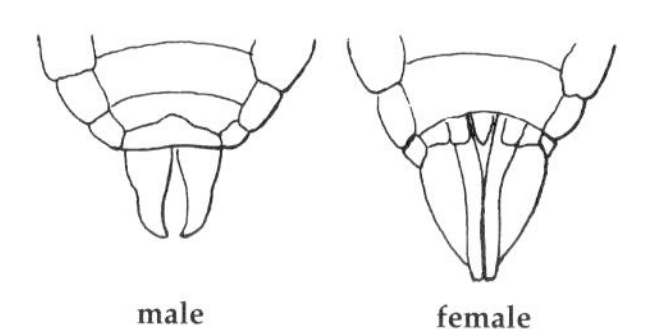

Fig. 15 Genitalia of *Philaenus spumarius*, seen from below.

The leaf beetle *Gastrophysa viridula* (Coleoptera: Chrysomelidae) (pls. 2, 4, 5, 6)

instar: one of several stages of development separated by moults

mesophyll: internal tissue of a leaf, responsible for much of the photosynthesis

pupa: immobile stage of insect development between larva and adult

This is perhaps the most characteristic of all the insects associated with *Rumex* plants. Adult *Gastrophysa* overwinter in soil. In mid April, the metallic green beetles emerge from hibernation and climb onto the *Rumex* plants, particularly *R. obtusifolius* and *R. crispus.* After a few days' feeding, the female beetles may be distinguished from males by their greatly distended black abdomens. Mating takes place and then batches of yellow eggs, usually 35 to 45 per batch, are deposited on the underside of leaves. There may be several batches on a single leaf. Eggs hatch after about one week and the black first-instar larvae stay in a clump close to where they have hatched and begin to chew the leaf surface to produce a fine net-like pattern (pl. 6), not usually penetrating below the upper mesophyll. After a further week or so, they moult to the second instar and these larger larvae disperse and chew actual holes in the leaf. By the third and last larval stage, damage to the leaf surface can be severe. Sometimes little but the veins remains. The third stage larvae (pl. 5) fall or climb off the plant and pupate in the soil to begin another generation. In favourable conditions the beetle may complete three generations, each of 5–6 weeks, before overwintering again. Adult beetles feed by chewing large holes in the leaves. They are easily dislodged from the plant but can climb back onto the same or a different plant. They are sometimes found in large numbers on the flowering stem of *Rumex* in amongst the developing flowers. The beautiful metallic colour of the adult beetles is not a pigment, but is the result of interference colours resulting from the structure of the cuticle. It can vary from green to almost copper-coloured.

A closely related species, *Galerucella nymphaeae* (pls. 2, 5), seems to be restricted to *Rumex hydrolapathum* where it replaces *G. viridula*. This is probably because *R. hydrolapathum* occurs in water-logged soil or standing in water. *G. viridula*, which pupates in the soil, is therefore unable to complete its life cycle, unlike *G. nymphaeae* which pupates attached to the leaves of the host plant.

Anthocorid bugs (pl. 1) and hoverfly larvae (pl. 5) are important predators of *Gastrophysa* eggs and larvae. Hoverflies lay their white eggs (usually singly) within batches of *Gastrophysa* eggs (pl. 4). When the hoverfly larva hatches, it eats the eggs or young larvae from its own batch and may then account for further egg batches nearby. It is not currently known which species of hoverfly attacks the first generation of *Gastrophysa* as the hoverfly larvae have proved difficult to rear through to adulthood in the laboratory for identification.

Some unanswered questions

The distribution of an insect species on different host plant species depends on the choices made by insects selecting plants on which to feed or lay their eggs, and also on their differential survival on the different plants. These possibilities can sometimes be separated by offering the insects a choice of host plant under controlled conditions in the laboratory to study behaviour, and examining the host-plant distribution of the different life history stages of the insects in the field to study survival. Which *Rumex* species do the beetles choose to lay their eggs or feed on if given a choice? How does this relate to their distribution in the field?

On other host plants, there is some evidence that plants respond to herbivore damage by some chemical change which may protect the leaf from further herbivory - or may make the leaf even more attractive to specialist herbivores. Do docks respond in this way? Do beetles avoid or prefer leaves which have already been damaged either by previous *Gastrophysa* feeding or by other insects such as *Philaenus spumarius* or *Pegomya*?

Does feeding by large numbers of beetles cause serious damage to the plant, for example by reducing seed production or root biomass? Is insect damage to the plants affected by the plants being subjected to drought? Information of this kind is needed for the planning of biological control programmes.

Mining flies, *Pegomya* species

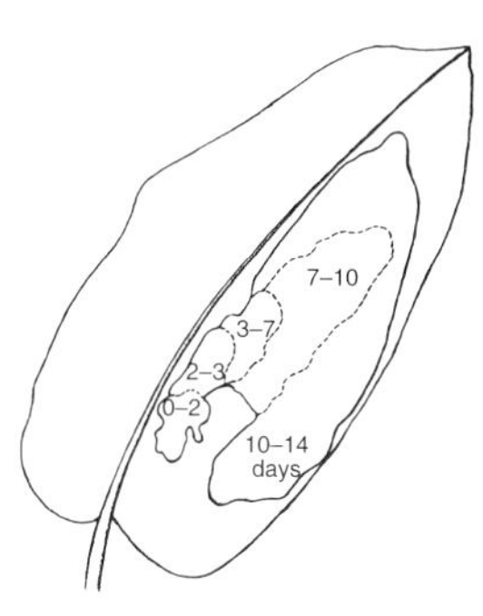

Fig. 16. Growth of *P. nigritarsis* leaf mine over two weeks in *R. obtusifolius* leaf (after Whittaker, 1992).

Anthomyiid flies, particularly *Pegomya nigritarsis* and *P. bicolor*, are specialist leaf miners of *Rumex* species. The flies lay their torpedo-shaped white eggs (3 x 0.5 mm) on the undersides of the leaves in small batches of one to about 10 eggs. The eggs are laid parallel to each other (pl. 4) and are easily distinguished from other eggs on the leaves, such as those of *Gastrophysa* (pl. 4). Eggs may be found from mid April to mid June (first generation) and again in September to October (second generation). After hatching the larvae burrow straight into the leaf and begin to feed beneath the upper epidermis to form a blotch mine (pls. 4, 6). Mines may contain one or several larvae. A large mine may completely cover the *Rumex* leaf, and if this happens before the larvae have completed their development, they may emerge from the mine and transfer to another leaf. In late spring, larval development is completed in about two weeks (fig. 16).

When larval growth is complete, the larvae leave the leaf and pupate in the soil or litter. *Pegomya* mines may be found on *R. obtusifolius, R. crispus, R. sanguineus, R. acetosa* and *R. acetosella*. The clutch size is roughly proportional to the leaf sizes in these host species (fig. 17), though this relationship is not a very strong one for different leaf sizes within a host species. Adult *Pegomya* are medium-sized flies (pl. 2).

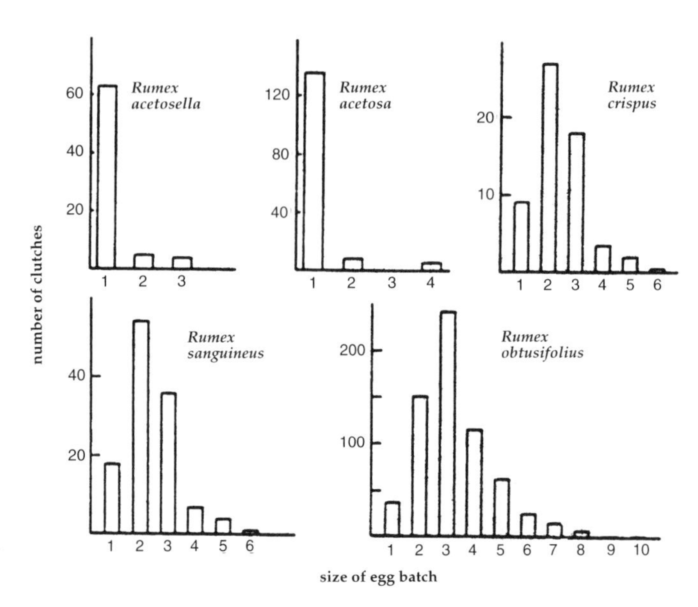

Fig. 17. Clutch size of *Pegomya* species on different *Rumex* species (Godfray, 1986).

There are at least 10 species of parasitic Hymenoptera associated with *Pegomya nigritarsis* on *R. obtusifolius* alone (table 2; Godfray, 1986). Some of these species are ectoparasites, which feed on, but outside, the body of the host. Others are endoparasites which feed inside the host's body. There is much to be done to work out the relationships between parasites and hosts.

Table 2. *Parasitoids of* Pegomya nigritarsis *on* Rumex obtusifolius *(adapted from Godfray, 1986).*

Parasite species	Family
Ectoparasites	
Colastes braconius	Braconidae
Chrysocharis nephereus	Eulophidae
Cirrospilus species	Eulophidae
Endoparasites	
Opius rufipes	Braconidae
Biosteres carbonarius	Braconidae
Biosteres impressus	Braconidae
Dapsilarthra florimela	Braconidae
Lamprotatus splendens	Pteromalidae
Skeleroceras truncatum	Pteromalidae
Trybliographa gracilicornis	Eucoilidae

As well as *Pegomya,* another fly often found on docks is *Norellisoma spinimanum* (pl. 2). Although the adults of this species are insectivorous, using bristles on their legs to catch prey, the larvae live within the leaf-stalks and stems of docks. Here the stem-boring larvae produce brown frass (excreta) which builds up around the entrance of the hole (pl. 4). The larvae can distort the stems and the holes are often easily spotted. Little is yet known about the parasitoids of these larvae although a pteromalid, *Seladerma breve* Walker, is recorded (Disney, 1976).

Some unanswered questions

Eggs of these leaf miners are attached to the leaf surface. They can be gently peeled off a leaf using the side of a fine needle, transferred to a different leaf, and re-attached with a tiny amount of a vegetable glue such as gum tragacanth. This procedure makes it possible to investigate a range of ecological questions. What is the optimum clutch size for *Pegomya*? Do larvae hatching from a single egg on a leaf of a given size have a higher likelihood of completing their development than larvae hatching from larger clutches? Do eggs laid on one species of *Rumex* have the same survival prospects if transferred to a different species? (An experiment designed to answer this question would require a control to test for the effect of disturbance alone; some eggs should be transferred back to their leaf of origin.) Does the rate of feeding of a larva depend on the host species, age of leaf, or nutritive status of the plant? This might be investigated by tracing mines onto a clear plastic sheet at 3 – 4 day intervals and measuring the area of the mine as it grows (fig. 16). Remember to allow for mines with different numbers of larvae in them.

Weevils (pls. 1, 4, 5)

With 12 to 14 species (Key V), weevils represent the biggest single assemblage of species on *Rumex*. The adult weevils are very mobile and may be observed on a wide range of *Rumex* species. To some extent the larvae can be distinguished on the basis of their host plants and feeding habits but there is much overlap between hosts and some overlap in feeding habits (fig. 8). The commonest species are probably *Apion violaceum* (pl. 1) and *A. curtirostre.* These are black weevils whose larvae are stem borers in several *Rumex* species. *Apion hydrolapathi* is difficult to distinguish from *A. violaceum,* but tends to occur in marshy areas, particularly in stems of *R. hydrolapathum. Apion rubens* attacks leaf stalks and mid-ribs of *R. acetosella.*

gall: a growth produced by a host in response to another organism

A number of *Apion* species are gall formers. The large red *Apion frumentarium* (= *miniatum*) attacks the base of the stem of *R. obtusifolius* and *R. crispus,* where it induces root galls which sometimes reach the size of a walnut. The larvae of this species are larger than other *Apion* species, the skin is rougher and there is usually a large cutting tooth on the left mandible. *A. sanguineum* causes galls in the roots and stems of *R. acetosella* (fig. 9). *A. rubens* larvae produce galls in the central vein or leaf stalk of *R. acetosella* and possibly other species, whilst *A. affine* and *A. cruentatum* form galls in the flower heads or stems and at the base of leaf stalks respectively in *R. acetosa*. *A. haematodes* larvae are found in the roots of *R. acetosella* and it is the commonest red weevil on this plant.

Other weevils which may be found on *Rumex* plants are *Rhinoncus perpendicularis* (pl. 1), and *R. pericarpius* (pl. 1), the larvae of which feed on roots of *R. obtusifolius* and *R. acetosa*. *Lixus* species may also pass their larval stages in *Rumex* stems. *Hypera rumicis* (pl. 1) is found on *R. crispus* and *R. obtusifolius*, though it also occurs on other Polygonaceae. Its eggs are laid in dark blister-like spots in the leaves (pl. 4). Here the larvae hatch, and then emerge to feed on the leaf (Chamberlin, 1933). Although superficially similar to *Gastrophysa* larvae, these larvae can be distinguished by their lines of small yellow spots along the back and sides. The fully-grown larva then spins a reddish-brown net-like cocoon on the plant or in nearby litter, in which it pupates (pl. 4).

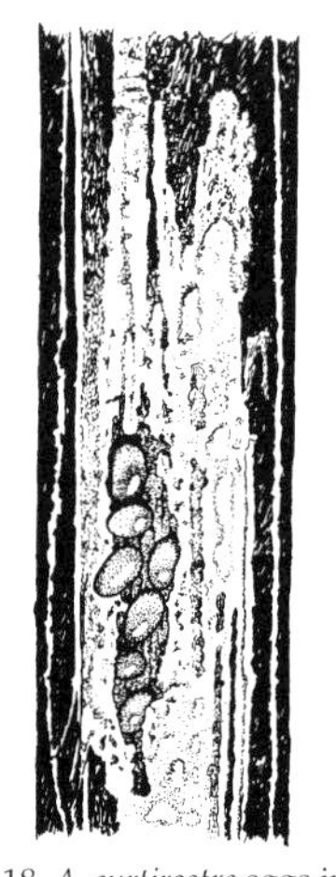

Fig. 18. *A. curtirostre* eggs in oviposition chamber in a cut-open *R. acetosa* stem.

Studies of *A. curtirostre* and *A. violaceum* have revealed much about life cycles of weevils on docks. Both these species have only one generation per year. The adults overwinter in leaf litter or on standing dead plants and in logs. They emerge in April or May to seek out *Rumex* hosts on which they feed and lay eggs in groups of up to 25 (mean 7.5) in the stem pith (fig. 18). The eggs are yellow when first laid, but they darken after one or two days and become black before hatching. In *R. acetosa*, the newly-hatched larvae bore their way through the thin layer of pith surrounding the central stem cavity. In *R. obtusifolius*, however, the pith of the stem does not decay so rapidly and *A. violaceum* larvae are found throughout the stem. There are three larval instars. Pupation takes place in a small cavity at the periphery of the stem. The adult emerges in June and July leaving a circular hole (pl. 4). Thus the time from egg to adult is approximately 5–6 weeks.

Parasites of weevils

Fig. 19. Exit hole left by weevil parasitoid in stem of *R. acetosa*.

There are at least four species of primary parasites, all Hymenoptera, which attack the range of weevil species found on *Rumex*. *Entedon rumicis* (pl. 1) and *E. hercyna* lay their eggs through the stem walls into the newly-laid eggs of *Apion* species. The parasite larva develops within the host until the host has made its exit tunnel. Then the host is killed and the parasite larva pupates. It overwinters as a pupa and emerges through the *Apion* exit tunnel the following spring leaving a neat round hole (fig. 19) which can easily be distinguished from those left by weevils (pl. 4). Weevils with ectoparasitoids tend to be found feeding higher up the host plant than weevils with endoparasitoids or unparasitised hosts (fig. 20). Parasitism is least frequent in weevils feeding in the thicker basal parts of the stem.

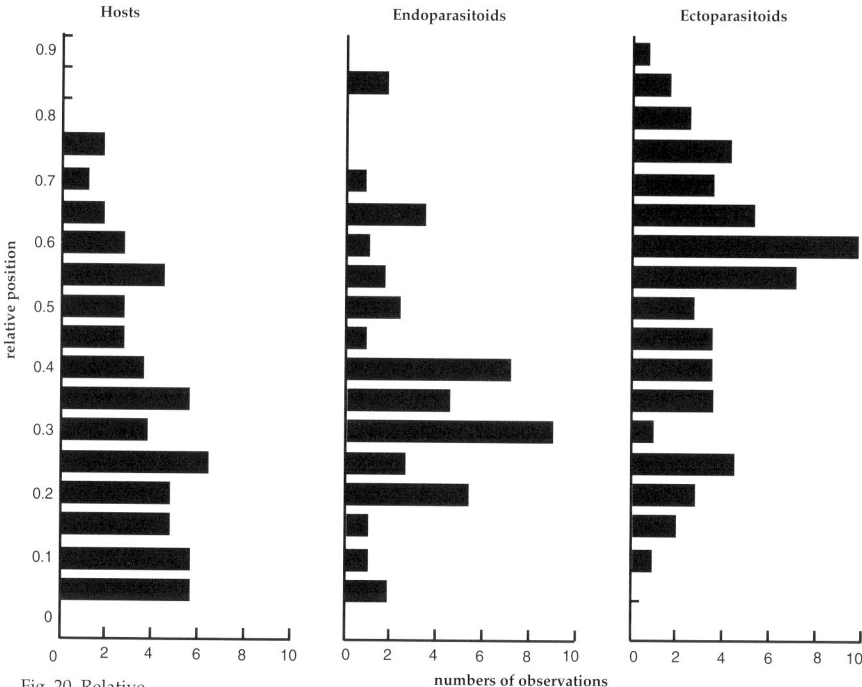

Fig. 20. Relative distributions of weevil hosts and ecto- and endoparasitoids from top (0.9) to bottom (0) of *R. crispus* stems (Freese, 1995).

Chlorocytus laogore and *Eurytoma curculionum* are ectoparasitic on *Apion* larvae. Again they utilise the exit tunnel of the host but whereas *C. laogore* pupates and emerges within weeks of the death of the host, *E. curculionum* larvae overwinter in the *Rumex* stem and pupate and emerge the following spring. In addition there are a number of hyperparasites that parasitise the primary parasites mentioned above. The whole parasite complex as at present understood (Hopkins, 1984) is illustrated in fig. 21, but the details of which parasites attack which hosts are still to be worked out. *Lucobracon erraticus* has recently been described as very common from *A. frumentarium* (= *miniatum*) in Europe (Freese, 1995).

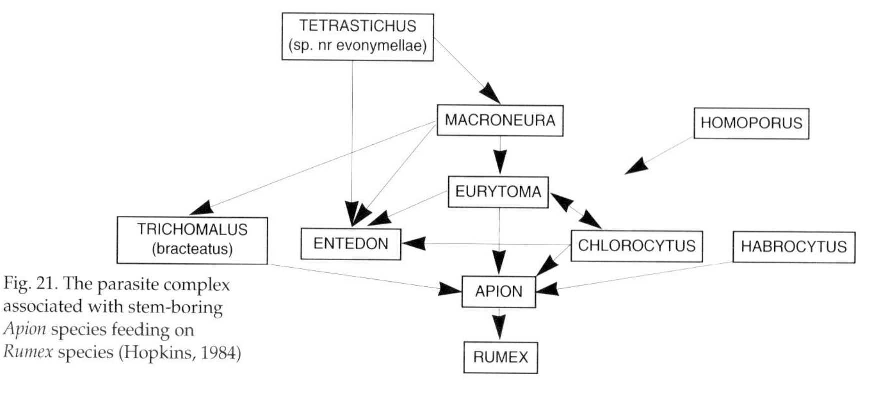

Fig. 21. The parasite complex associated with stem-boring *Apion* species feeding on *Rumex* species (Hopkins, 1984)

Some unanswered questions

Evidence of attack by these weevils is visible in the form of emergence holes in *Rumex* stems. The larger (about 1 mm diameter) holes have an elliptical depression around them and are exit points of the adult weevils. Smaller holes (less than 1 mm) without the depression show where parasites have emerged. *Rumex* stems, especially of the more robust species such as *R. obtusifolius*, remain standing long after death. It is therefore possible to record densities (numbers of emergence holes per stem) at the end of the current season, but also to find evidence of densities in the preceding year(s) if the dead stems remain. Detective work of this kind makes it possible to investigate how the weevils are distributed between species of *Rumex* and within the plant. Do they tend to emerge near to the base of a stem or at random along its length? What are the rates of parasitism of the weevils at different sites on different *Rumex* species? How much do populations of weevils, and rates of parasitism, vary between years?

Copper butterflies

There are two native species of copper butterflies (family Lycaenidae) in the British Isles, the small copper *Lycaena phlaeas* and the large copper *Lycaena dispar*. They are among the most brilliantly coloured butterflies in the British fauna.

Small copper

This species lays its eggs singly on the undersides of smaller dock species such as sorrel and sheep's sorrel, although it occasionally uses larger species. The larva feeds in a groove it excavates beneath the leaf which is visible as a window from above. There are normally two generations per year but occasionally there are three or four depending on summer weather. Larvae from both summer and autumn broods overwinter on the food plant and complete their development the following year. The pupal stage lasts up to a month after which the butterflies emerge. Egg-laying behaviour is greatly influenced by weather; females will only lay in sunny conditions and they place their eggs on the plants in the brightest sunshine. The insect is widely distributed in the British Isles (fig. 22).

Large copper

The large copper had a restricted distribution in the British Isles, in the fens of East Anglia and probably in Somerset and Monmouthshire. The British subspecies of this butterfly *Lycaena dispar dispar* became extinct around 1865, perhaps because of fen drainage and over-collecting. Since that time attempts have been made to reintroduce the European subspecies *Lycaena dispar batavus* and *rutilus*. The former subspecies still persists in a protected colony dating from 1928 at Woodwalton Fen in Cambridgeshire. Therefore, the distribution of this species (fig. 23) is very different from the wider distribution of the small copper. It is also becoming increasingly rare in mainland Europe. Eggs are laid singly on leaves of the great water dock, with up to five eggs on each leaf. In captivity other *Rumex* species are accepted by the larvae. The species has only one generation per year.

Decomposers and other invertebrates

If dock stems containing cavities made by fly or weevil larvae are dissected, other groups of invertebrates may be found. Some of these may be predators or parasites of the herbivores. Some may be living on decaying matter in the cavities such as dead plant material, herbivore frass, or on the fungi and bacteria associated with these. Others may simply be sheltering in the cavity. Some of these invertebrates, such as springtails, molluscs and millipedes, eat a wide range of food material. These can be keyed out in Key II.

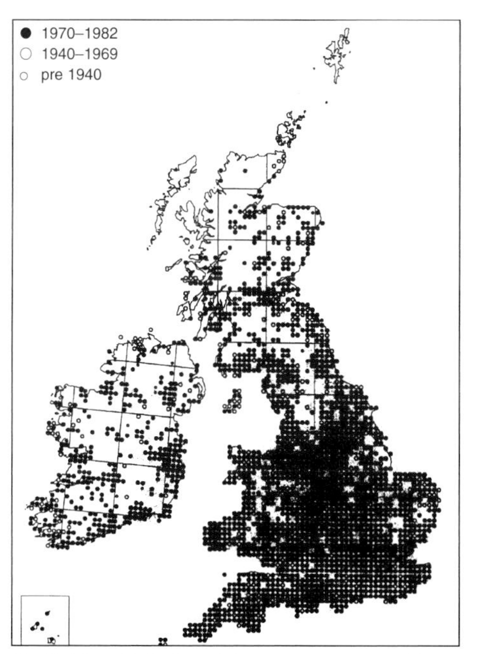

Fig. 22. The distribution of the small copper butterfly in the British Isles.

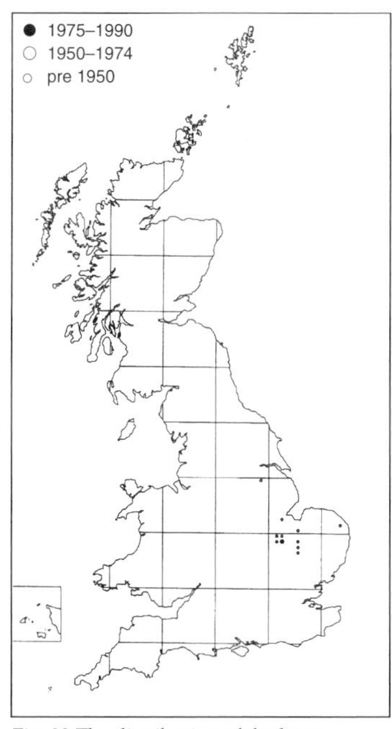

Fig. 23 The distribution of the large copper butterfly in the British Isles.

4 Identification

Introduction to the keys

These keys are designed for the identification of docks and sorrels and the insects associated with them. They will also identify, to group level, 'tourist' insects, insects which have no special relationship with docks but may fall onto them from overhanging trees and bushes, be blown onto them, or simply alight on the leaves. The keys are designed to help with the identification of immature stages as well as adults. Unfortunately the taxonomy of juvenile insects is still poorly known and for many groups no keys to them are available. Unfamiliar technical terms are explained in illustrations that accompany each key.

I Docks and sorrels

Eleven of the important species of *Rumex* from a variety of habitats are separated using this key. The key is designed for mature plants, and most of the characters used are based on the inflorescence. A x10 magnification hand lens and ruler graduated in mm will be useful.

A *Rumex* flower consists of six greenish bract-like structures which are not differentiated into petals and sepals, and are therefore called perianth segments (or, sometimes, tepals). At fruiting, the inner three perianth segments can develop on the surface a swelling or tubercle, the shape of which is diagnostic.

Several species, however, can be distinguished by their general growth pattern and habitat. Leaf shape should be determined from basal leaves, as these are often different from those on the stem, although they may have died away on mature plants. Hybrids may have intermediate characteristics but generally grow in the vicinity of the more easily identified parent plants. Seedlings can be recognised as docks at the two cotyledon stage by their thin-stalked cotyledons (fig. 24) (Chancellor, 1959). A valuable comprehensive key to British docks and knotweeds is given by Lousley & Kent (1981).

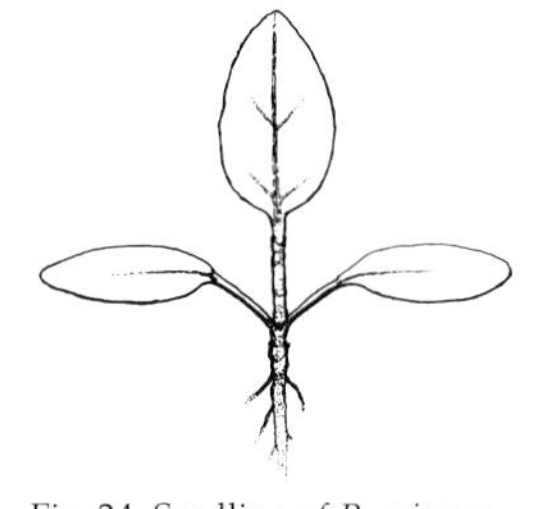

Fig. 24. Seedling of *R. crispus.*

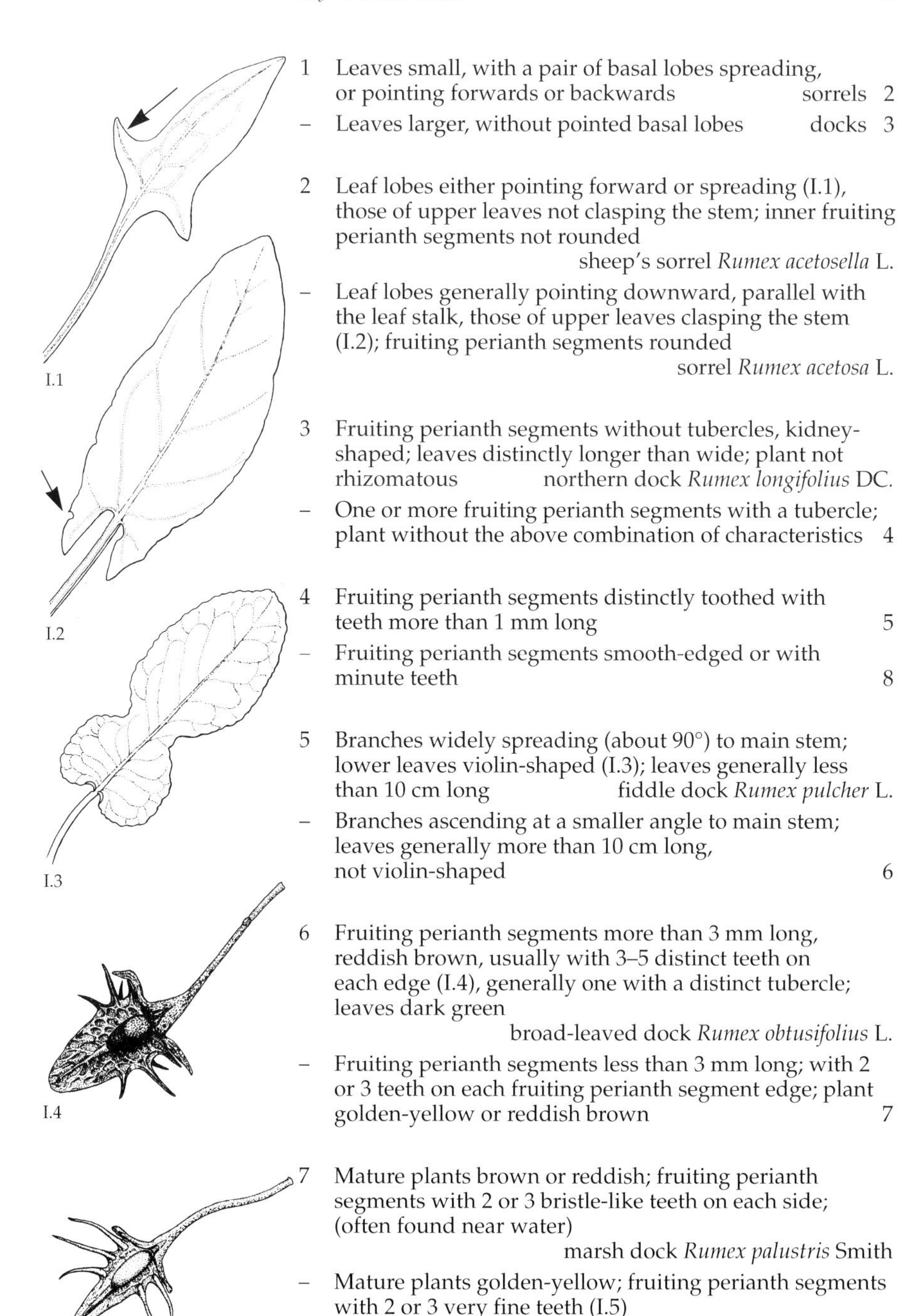

1 Leaves small, with a pair of basal lobes spreading, or pointing forwards or backwards — sorrels 2

– Leaves larger, without pointed basal lobes — docks 3

2 Leaf lobes either pointing forward or spreading (I.1), those of upper leaves not clasping the stem; inner fruiting perianth segments not rounded
sheep's sorrel *Rumex acetosella* L.

– Leaf lobes generally pointing downward, parallel with the leaf stalk, those of upper leaves clasping the stem (I.2); fruiting perianth segments rounded
sorrel *Rumex acetosa* L.

3 Fruiting perianth segments without tubercles, kidney-shaped; leaves distinctly longer than wide; plant not rhizomatous — northern dock *Rumex longifolius* DC.

– One or more fruiting perianth segments with a tubercle; plant without the above combination of characteristics — 4

4 Fruiting perianth segments distinctly toothed with teeth more than 1 mm long — 5

– Fruiting perianth segments smooth-edged or with minute teeth — 8

5 Branches widely spreading (about 90°) to main stem; lower leaves violin-shaped (I.3); leaves generally less than 10 cm long — fiddle dock *Rumex pulcher* L.

– Branches ascending at a smaller angle to main stem; leaves generally more than 10 cm long, not violin-shaped — 6

6 Fruiting perianth segments more than 3 mm long, reddish brown, usually with 3–5 distinct teeth on each edge (I.4), generally one with a distinct tubercle; leaves dark green
broad-leaved dock *Rumex obtusifolius* L.

– Fruiting perianth segments less than 3 mm long; with 2 or 3 teeth on each fruiting perianth segment edge; plant golden-yellow or reddish brown — 7

7 Mature plants brown or reddish; fruiting perianth segments with 2 or 3 bristle-like teeth on each side; (often found near water)
marsh dock *Rumex palustris* Smith

– Mature plants golden-yellow; fruiting perianth segments with 2 or 3 very fine teeth (I.5)
golden dock *Rumex maritimus* L.

8 Inflorescence diffuse; fruiting perianth segments usually not more than 3 mm long — 9

– Inflorescence dense; fruiting perianth segments usually at least 3 mm long; tubercles not more than half as long as fruiting perianth segments — 10

I.6

9 Stems almost straight, branches ascending at about 20°; only one, globular tubercle on one fruiting perianth segment (I.6) — wood dock *Rumex sanguineus* L.

– Stems usually wavy, branches at 30–90°; each fruiting perianth segment with a single oblong tubercle; (generally in wet habitats) — clustered dock, sharp dock *Rumex conglomeratus* Murray

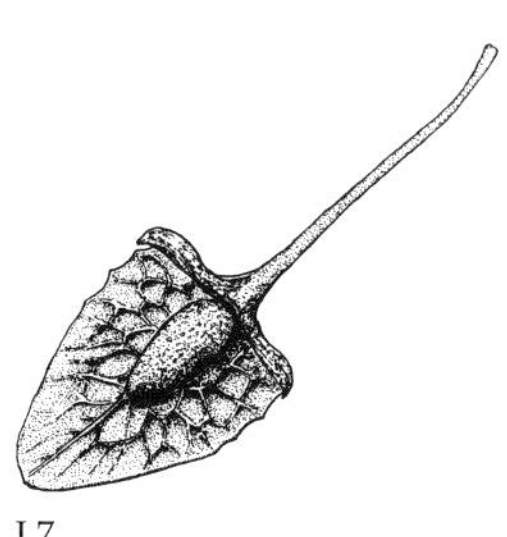

I.7

10 Plant up to 2 m tall; leaves without wavy or curled edges, dull green; fruiting perianth segments smooth-edged or with small teeth near base; seed at least 3 mm long; inflorescence dense; (wet habitats) — great water dock *Rumex hydrolapathum* Hudson

– Plant smaller; leaves with wavy or curled edges, narrowly lanceolate; fruiting perianth segments smooth-edged or with minute teeth on the edges (I.7); seed not more than 3 mm long — curled dock *Rumex crispus* L.

II Invertebrate groups

This key (adapted from Paviour-Smith & Whittaker, 1967) separates the main groups of insects, and shows which of the later keys (Keys III–VII) to use for further identification. It includes insects associated directly with *Rumex* species through feeding, and the commoner specialist parasitoids or predators of these insects. Parasitoids are keyed as far as practicable. Some of the visitors or decomposer organisms which may be found on these plants are also included, but no attempt is made to key these out beyond main groups.

These keys must be used with care. If other insects turn up, not known to be associated with *Rumex*, they may be missing from the key (in which case a book such as Chinery (1986) can give a useful start), or they may key out in the wrong place. It is therefore essential to check specimens in detail against the descriptions and illustrations in this book, and if necessary against those in the more comprehensive, specialist keys appropriate for each group.

Names of species are given in italics, those of phyla and orders in capital letters and those of other taxa such as families (which end in -idae) in lower case with a capital initial letter.

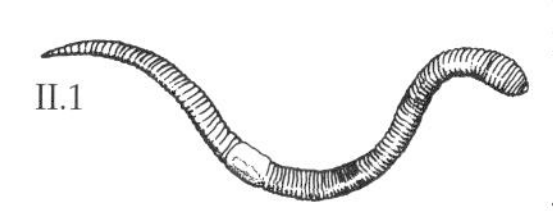
II.1

1 Body not segmented — 2
– Body with segments which are visible (II.1), or indicated by the presence of paired, jointed limbs — 3

II.2

2 Animals soft-bodied and slimy; lower surface of body forming a muscular foot (II.2) — slugs and snails (MOLLUSCA)
– Animals with a tough shining cuticle; small, worm-like — threadworms (NEMATODA)

3 Without jointed legs — 4
– With jointed legs — 6

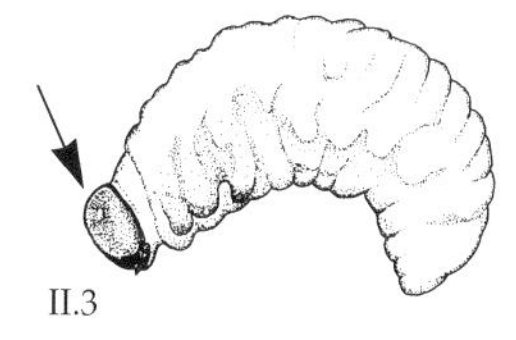
II.3

4 body segments many more than 15 (no head capsule) — worms (ANNELIDA)
– Body with fewer than 15 segments (exclusive of any obvious head) — certain insect larvae 5

5 With a distinct head capsule (II.3) — weevil larvae (COLEOPTERA, Curculionidae (part)) Key V
– Without a distinct head capsule (II.4); mouthparts reduced, hard and dark — larvae of higher flies (Diptera) Key VII

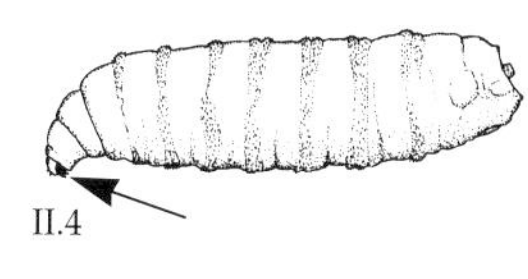
II.4

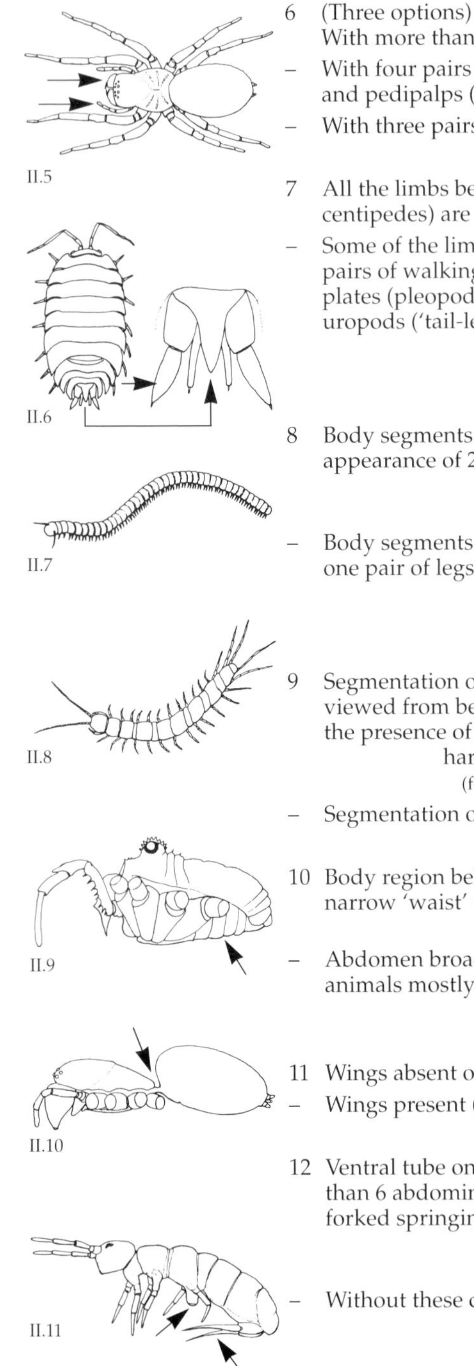

II.5

II.6

II.7

II.8

II.9

II.10

II.11

6 (Three options)
With more than four pairs of jointed legs 7
– With four pairs of jointed legs, and with chelicerae and pedipalps (II.5) ARACHNIDA 9
– With three pairs of jointed legs INSECTA 11

7 All the limbs behind head (except the first pair in centipedes) are walking legs 8
– Some of the limbs behind head are not walking legs (6–7 pairs of walking legs are followed by 5 pairs of small flat plates (pleopods; visible from below) and 1 pair of uropods ('tail-legs') (II.6)
woodlice (ISOPODA, CRUSTACEA)
(for further identification see Hopkin, 1991)

8 Body segments (after first 3) fused in pairs, giving appearance of 2 pairs of legs per 'segment' (II.7)
millipedes (DIPLOPODA)
(for further identification see Blower, 1958)
– Body segments not fused in pairs, so that there is only one pair of legs per apparent segment (II.8)
centipedes (CHILOPODA)
(for further identification see Eason, 1964)

9 Segmentation of abdomen distinct (II.9) (may need to be viewed from below; do not confuse segmentation with the presence of a 'waist', II.10)
harvestmen (OPILIONES or PHALANGIDA)
(for further identification see Hillyard & Sankey, 1989)
– Segmentation of abdomen not visible externally 10

10 Body region behind the legs separated by a narrow 'waist' (II.10) spiders (ARANEIDA)
(for further identification see Roberts, 1995)
– Abdomen broadly fused to rest of body; animals mostly minute, some small mites (ACARINA)
(for further identification see Evans, 1992)

11 Wings absent or very small (wing buds) 12
– Wings present (but may be folded under wing cases) 18

12 Ventral tube on first abdominal segment (II.11); not more than 6 abdominal segments (the fourth often bearing a forked springing organ); animals small or minute (II.11)
springtails (COLLEMBOLA)
(for further identification see Fjellberg, 1980)
– Without these characters 13

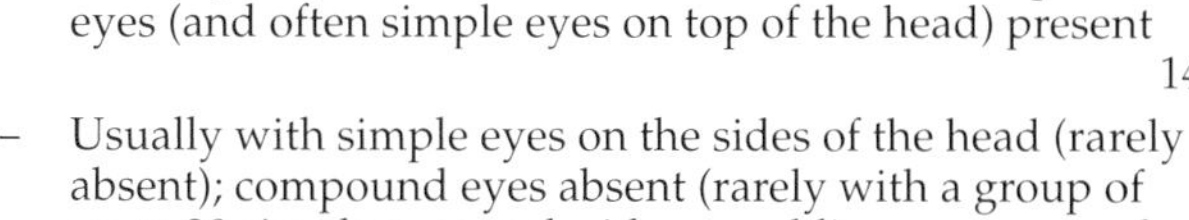

13 No simple eyes on the sides of the head, but compound eyes (and often simple eyes on top of the head) present 14

– Usually with simple eyes on the sides of the head (rarely absent); compound eyes absent (rarely with a group of up to 30 simple eyes each side resembling a compound eye, but if so the animal is very pale and soft) remaining larvae of higher insects 15

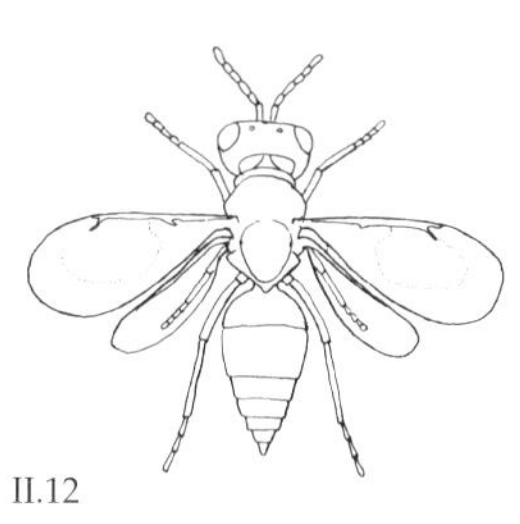

II.12

14 Mouthparts modified as long piercing and sucking stylets (these insects usually have a triangular-shaped face) (II.16) nymphal and wingless bugs (HEMIPTERA) Key III

– Mouthparts present but not modified as stylets; with a narrow waist (as in II.12) ants (HYMENOPTERA) Key VI

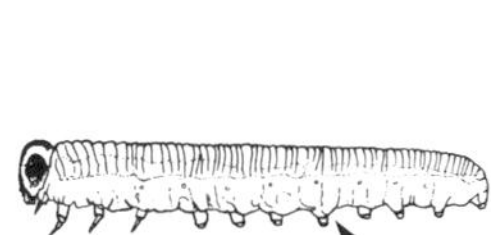

II.13

15 With pairs of unjointed false legs (prolegs (II.13)) on abdomen as well as jointed legs on thorax 16

– Without pairs of prolegs on abdomen 17

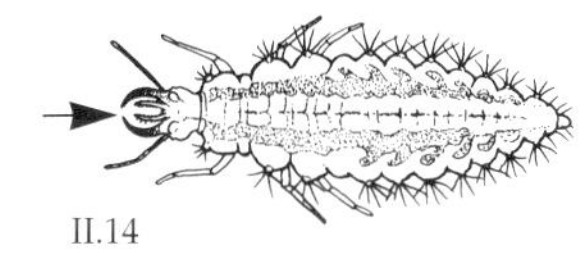

II.14

16 Prolegs bearing rings or rows of minute hooks at the tip caterpillars (larvae of LEPIDOPTERA) Key IV

– Prolegs without such hooks (II.13) sawfly larvae (larvae of SYMPHYTA, HYMENOPTERA)
The common sawfly larva on *Rumex* is included with the LEPIDOPTERA larvae in Key IV.

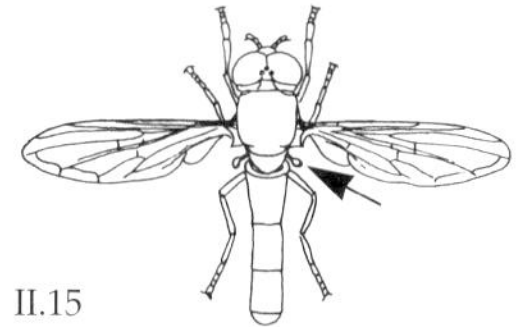

II.15

17 Larvae parallel-sided even if tapering at ends beetle larvae (COLEOPTERA) Key V

– Larvae spindle-shaped (II.14) with very large and obvious jaws larvae of lacewings (NEUROPTERA) (for further identification see Plant, 1997)

18 Only one pair of wings, the other pair modified as 'halteres' (small club-like organs) (II.15) adult flies (DIPTERA) Key VII

– Two pairs of wings (but in beetles the forewings (elytra) are hardened and may be fused along the mid line) 19

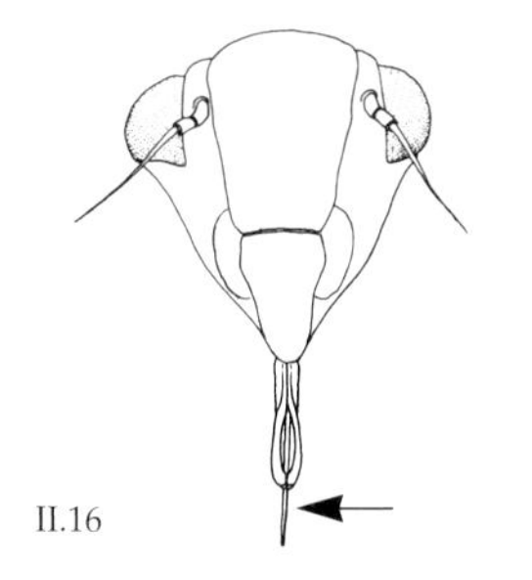

II.16

19 Mouthparts modified as long sucking stylets; face usually triangular-shaped (II.16) bugs (HEMIPTERA) Key III

– Mouthparts not like this 20

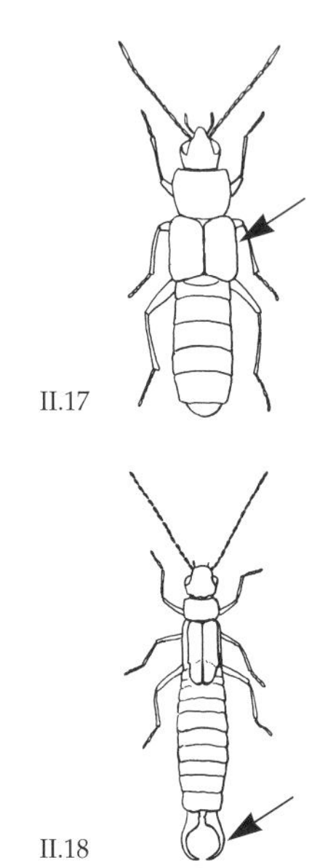

II.17

II.18

20 Forewings forming usually hard wing covers (elytra (V.4)) which usually meet in the mid-dorsal line without overlapping; membranous wings folded longways and crossways under elytra — 21

– Forewings at least partly membranous, often overlapping; hindwings may be folded longways but are never folded crossways under elytra — 23

21 Elytra covering most of the abdomen — adult beetles (COLEOPTERA) Key V

– Elytra always short exposing most of abdomen (II.17) — 22

22 Abdomen ending in a pair of unjointed, more or less converging 'forceps' (II.18) — earwigs (DERMAPTERA)
(for further identification see Chinery, 1993)

– Abdomen without forceps at the end (II.17), though may have more slender terminal style — adult beetles (COLEOPTERA) Key V

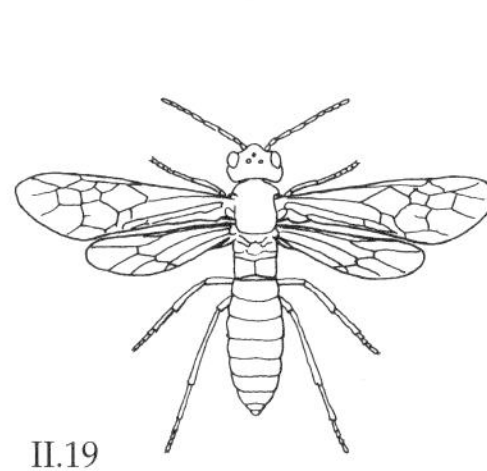

II.19

23 Wings and body covered with flattened scales; mouthparts, if present, a coiled proboscis (tongue) — moths and butterflies (LEPIDOPTERA) Key IV

– Wings without flattened scales, but they may bear hairs; forewings interlocking with hindwings by means of a ridge on the former and row of hooks on the latter; sometimes with very much reduced venation, elbowed antennae, and/or a dark spot on leading edge of forewing) (II.19) — HYMENOPTERA Key VI

PLATE 1

1. *Hypera rumicis* (x 7)
2. *Apion violaceum* (x 7)
3. *Apion frumentarium* (= *miniatum*) (x 7)
4. *Rhinoncus pericarpius* (x 7)
5. *Rhinoncus perpendicularis* (x 7)
6. *Anthocoris nemorum* (x 7)
7. *Entedon rumicis* (x 9)
8. *Contarinia acetosellae* (x 9)
9. *Aphis rumicis* (x 9)

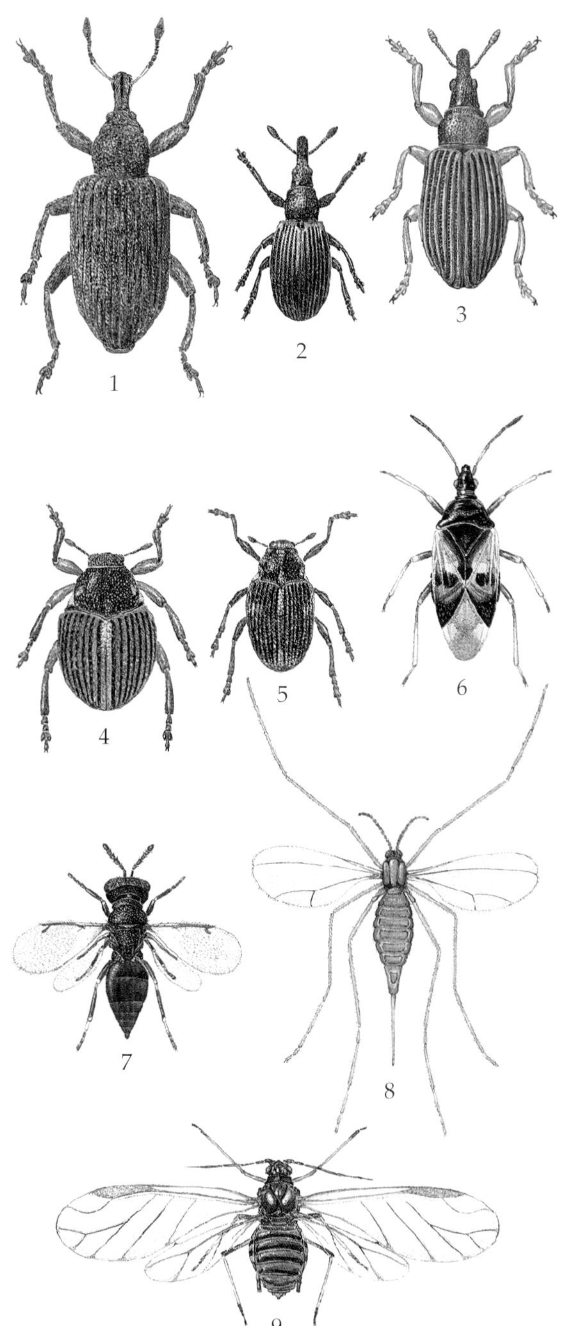

PLATE 2

(All x 3.5)

1. *Gastrophysa viridula*
2. *Galerucella nymphaeae*
3. *Metasyrphus luniger*
4. *Platycheirus scutatus*
5. *Norellisoma spinimanum*
6. *Pegomya bicolor*
7. *Pegomya nigritarsis*
8. *Ametastegia glabrata*
9. *Philaenus spumarius*

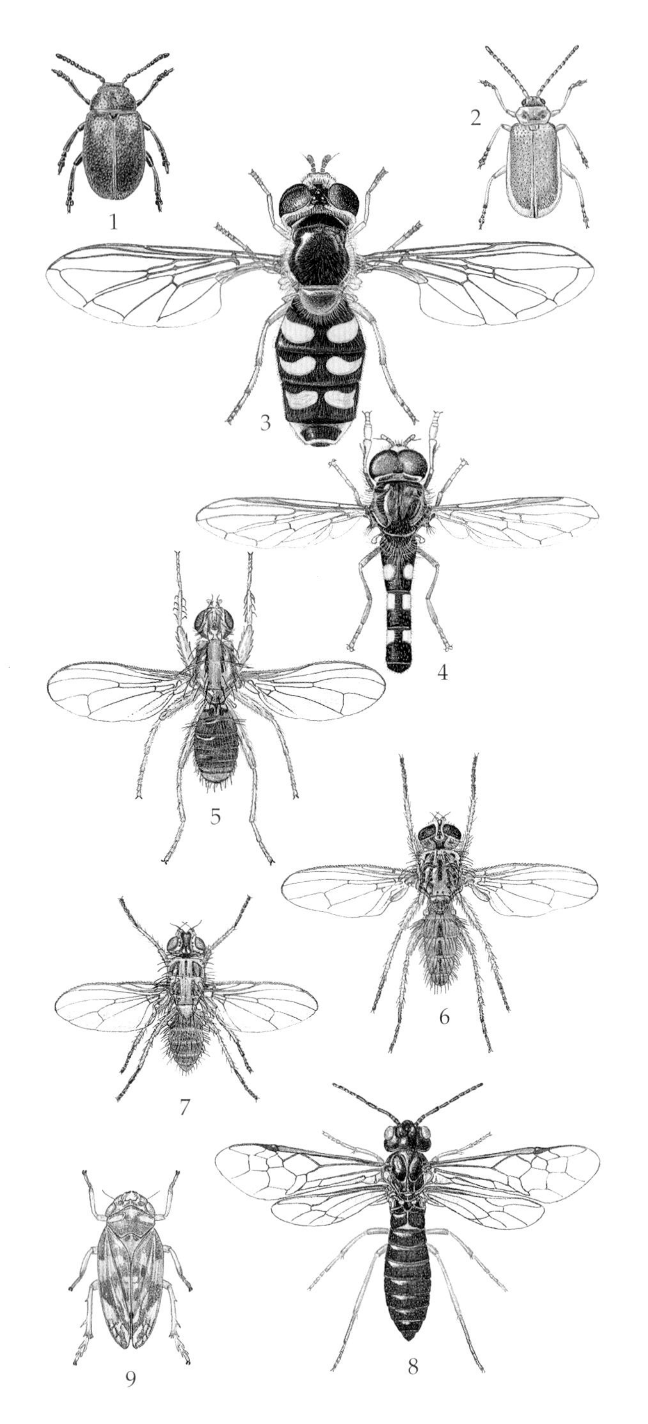

PLATE 3

1. Garden tiger
   Arctia caja (x 1.5)
2. Angle shades
   Phlogophora meticulosa (x 1.5)
3. Buff ermine
   Spilosoma lutea (x 1.5)
4. Bright-line brown-eye
   Lacanobia oleracea (x 1.5)
5. Heart and dart
   Agrotis exclamationis (x 1.5)
6. Silver Y
   Autographa gamma (x 1.5)
7. Large copper
   Lycaena dispar (x 3)
8. Cyclamen tortrix
   Clepsis spectrana (x 3)
9. Sawfly larva
   Ametastegia glabrata (x 3)

1

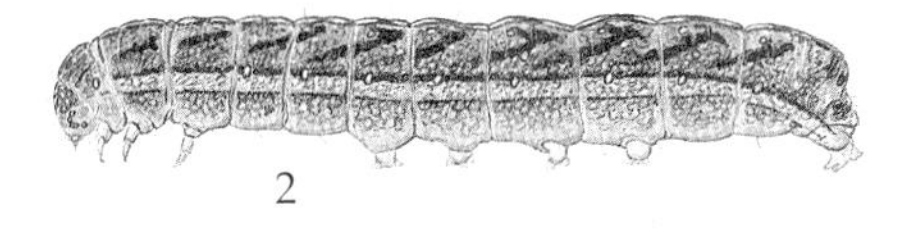

2

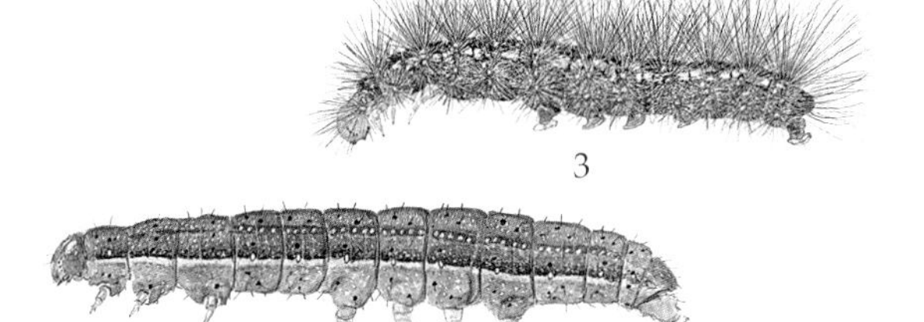

3

4

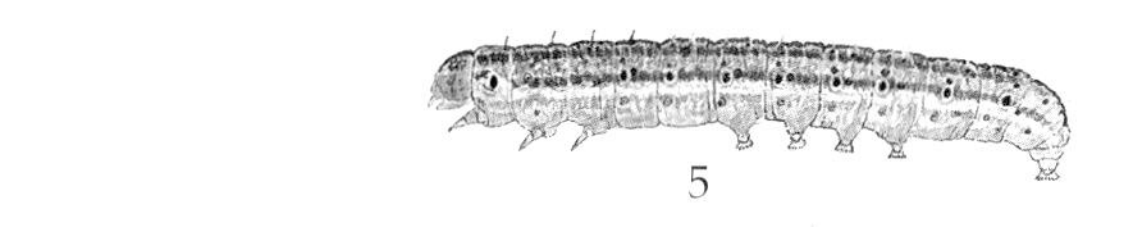

5

6

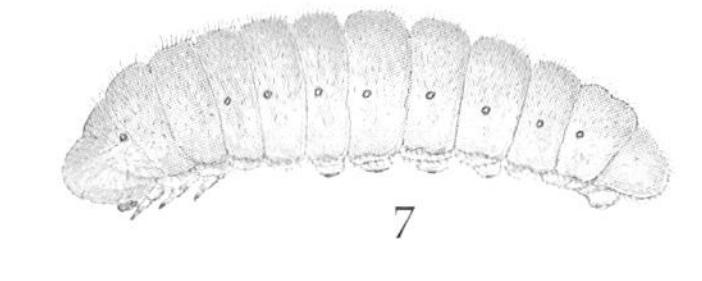

7

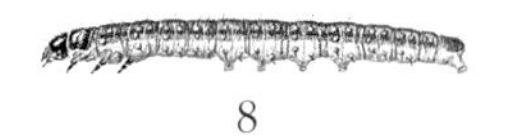

8

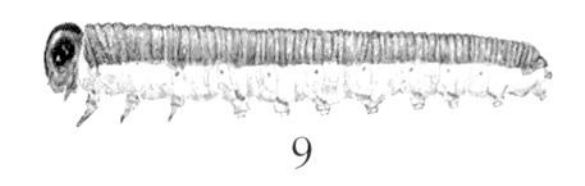

9

PLATE 4

1. *Hypera rumicis* leaf blisters (x 3)
2. *Hypera rumicis* cocoon (x 3)
3. *Gastrophysa* egg clutch with syrphid egg in centre (x 3)
4. Adult emergence hole of *Apion violaceum* (x 6)
5. *Olethreutes lacunana* leaf roll (x 2)
6. Larval boring of *Norellisoma spinimanum* (x 6)
7. *Pegomya* egg clutch (x 3)
8. Blotch mine of *Pegomya* (x 1)

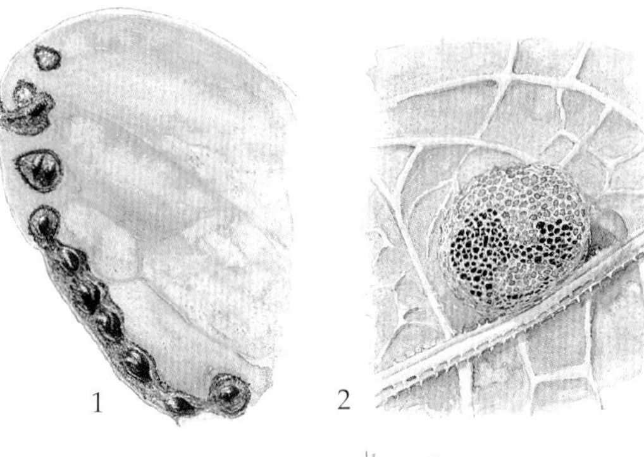

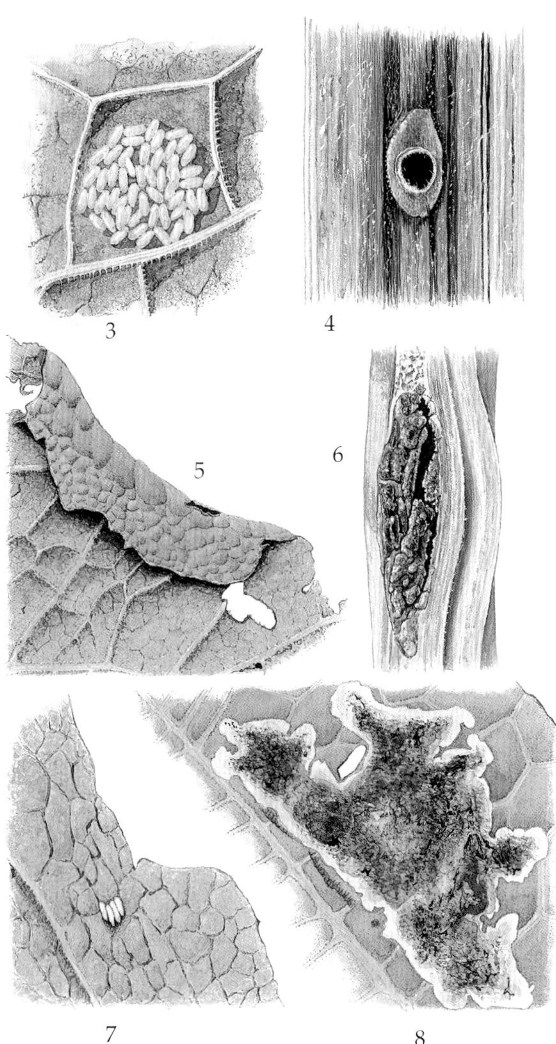

PLATE 5

(All x 7)

1. *Gastrophysa viridula* larva
2. *Galerucella nymphaeae* larva
3. *Galerucella nymphaeae* pupa
4. *Hypera rumicis* larva
5. *Philaenus spumarius* nymph
6. *Apion frumentarium* (= *miniatum*) larva
7. *Pegomya* larva
8. *Norellisoma spinimanum* larva
9. *Syrphus vitripennis* larva

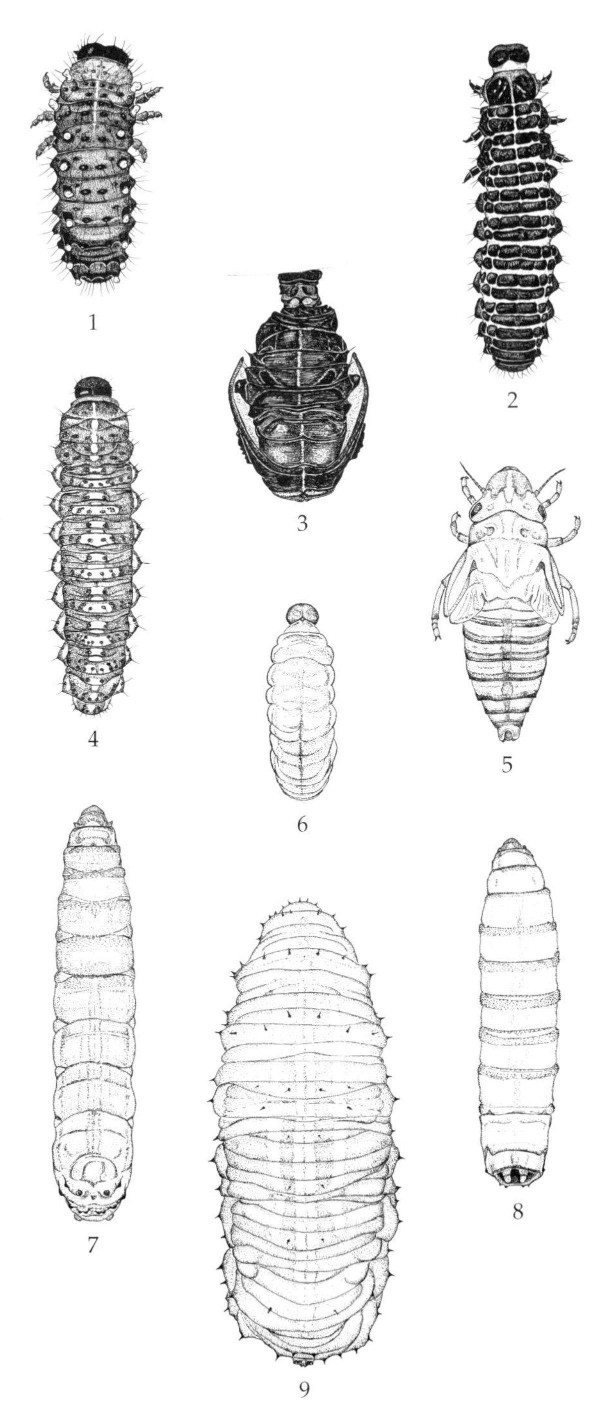

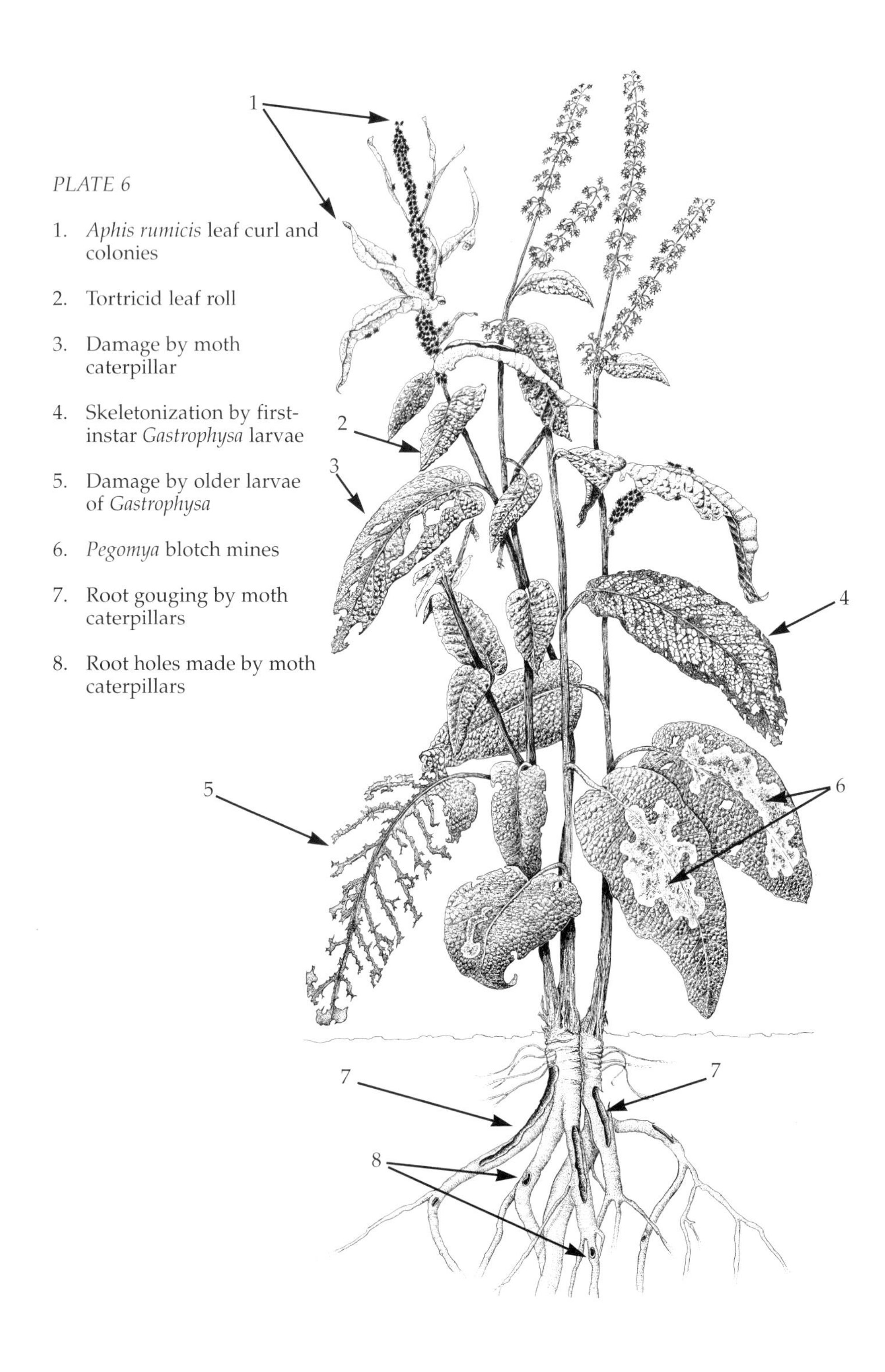
PLATE 6
1. Aphis rumicis leaf curl and colonies
2. Tortricid leaf roll
3. Damage by moth caterpillar
4. Skeletonization by first-instar Gastrophysa larvae
5. Damage by older larvae of Gastrophysa
6. Pegomya blotch mines
7. Root gouging by moth caterpillars
8. Root holes made by moth caterpillars
1
2
3
4
5
6
7
7
8

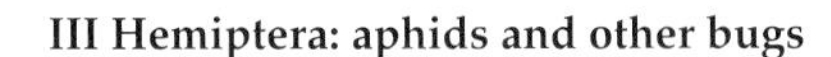

III Hemiptera: aphids and other bugs

The key to aphids has been adapted from Stroyan (1984). Some aphid species on docks are attended by ants (Skinner & Allen, 1996).

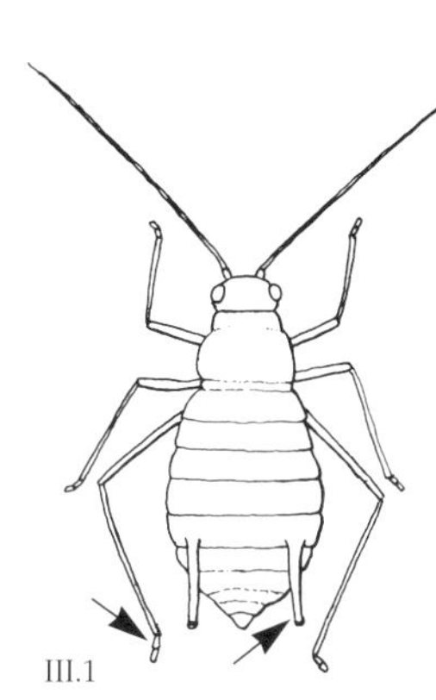

III.1

1 Wingless — 2
– Winged — 8

2 With a pair of cornicles (III.1) on top of fifth abdominal segment; antennae long, threadlike; tarsi 2-segmented wingless and immature aphids — 3
– Without cornicles on abdomen; tarsi 3-segmented — 7

3 Aphids feeding deep in soil on roots, or just below soil surface, or at root collar — 4
– Aphids feeding on shoot only — 5

4 Aphids just below the soil surface; usually ant-attended; dark green-brown or blackish; on *R. acetosa* or *R. acetosella*; rare — *Aphis acetosae* L.
– Aphids found deeper in soil — 6

5 Aphids black or nearly black — 13
– Aphids yellow to green — 14

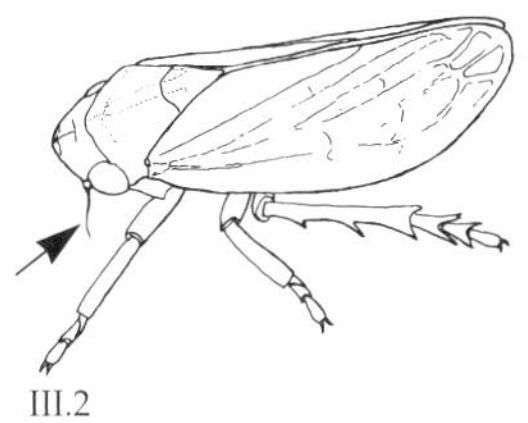

III.2

6 Aphids bluish-grey — *Dysaphis radicola* (Mordvilko)
– Aphids pale yellow-whitish, or bluish-green, living on deep roots — 15

7 Antennae shorter than head (III.2); body not dorso-ventrally flattened (pl. 5) spittlebug *Philaenus spumarius* (L.)
– Antennae longer than head nymphs of true bugs (HETEROPTERA)
(for further identification see Southwood & Leston, 1959)

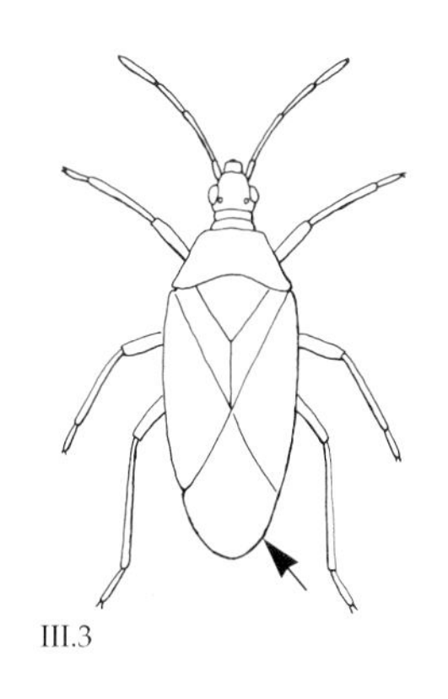

III.3

8 Forewings generally of uniform consistency all over; wings usually held sloping in tent-like fashion over sides of body (III.2) — HOMOPTERA 9
– Forewings with a differently-textured region at the tip, when folded lying flat on the back, with tips overlapping (III.3) — HETEROPTERA 10

9 Antennae short, ending in a bristle; forewings opaque, tarsi with 3 segments (pl. 2)
froghoppers (Auchenorrhyncha) *Philaenus spumarius*

– Antennae well developed, without a conspicuous bristle at the end; forewings transparent, tarsi with 1 or 2 segments winged aphids (Sternorrhyncha) 13

10 Simple eyes absent Miridae 11

– Simple eyes present 12

11 Head and thorax green *Lygocoris pabulinus* (L.)

– Head and thorax largely brown or black
Lygus rugulipennis Poppius

A generalist species, recorded from nearly 400 host species

12 Beak-like mouthparts (rostrum) curved in side view, body more than 6 mm long Nabidae

A family of predatory bugs which may be encountered on docks

– Rostrum not curved, less than 6 mm long
Anthocoridae *Anthocoris* species
probably *A. nemorum* (L.) (pl. 1)

13 Third and most of fourth antennal segment pale; body often with white wax patches on back; not causing leaf curling bean aphid *Aphis fabae* (Scop.)

– Most of third antennal segment and all of fourth segment dark; body without white wax patches; causing leaf curling; often attended by ants
permanent dock aphid *Aphis rumicis* L. (pl. 1)

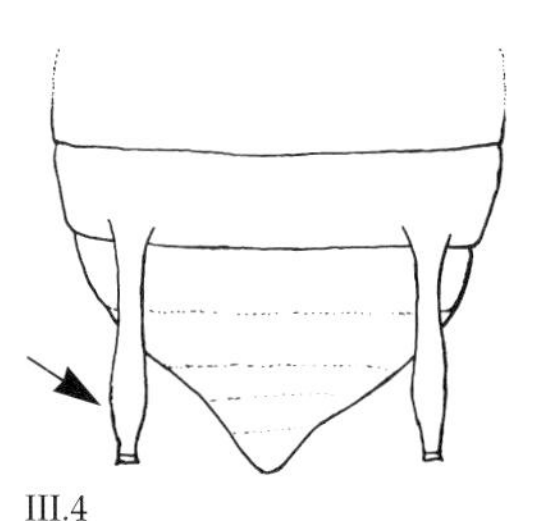
III.4

14 Cornicles with a bulbous widening near the end and slightly flared at the tip (III.4); aphids light yellow to green peach-potato aphid *Myzus persicae* (Sulzer)

– Cornicles without bulbous widening or flared tips
Myzus ornatus Laing
Aulacorthum solani (Kaltenbach)
Macrosiphum euphorbiae (Thomas)

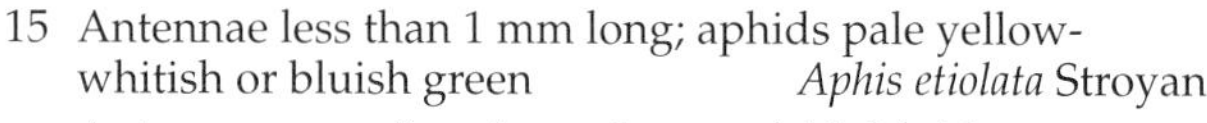
15 Antennae less than 1 mm long; aphids pale yellow-whitish or bluish green *Aphis etiolata* Stroyan

– Antennae more than 1 mm long; aphids bluish-green
Aphis sambuci L.

IV Lepidoptera: adult and larval moths and butterflies

Adults

Adult Lepidoptera are difficult to associate with food plants unless the adults are reared from larvae feeding on the plants. Moths may shelter and butterflies may sun themselves on dock leaves with which they have no other association. So particular care is needed in using this key to name adults. Many of the butterfly and moth species in these keys are illustrated in Carter & Hargreaves (1986) and Thomas & Lewington (1991).

Wingspan, used in this key as a general guide, is measured between the furthest tips of the outstretched forewings in set specimens.

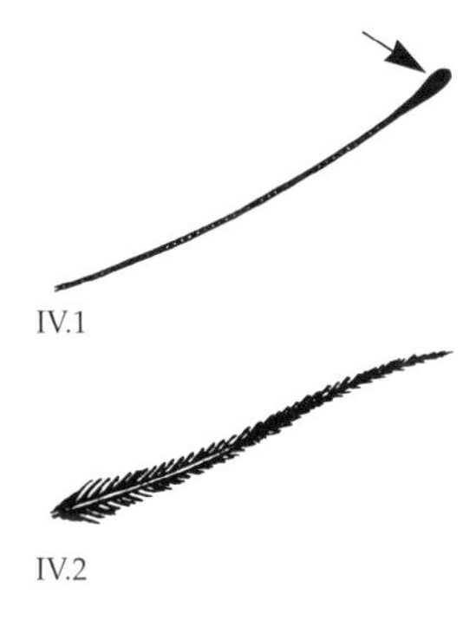

IV.1

IV.2

1 Antennae long, ending in clubs (IV.1)
butterflies (only Lycaenidae are associated with docks) 2

– Antennae short and feathered or tapering (IV.2)
moths 3

Butterflies (Lycaenidae)

2 Bright coppery wings without small projections or 'tails' on the hindwing; wingspan
39–50 mm large copper *Lycaena dispar* Haworth

– Bright coppery wings with 'tails' on the margin of hind wing, wingspan 23–28 mm
small copper *Lycaena phlaeas* L.

Moths

3 Brightly coloured and stout moths 4

– Cryptically coloured and stout or flimsy-looking, thin-bodied moths 8

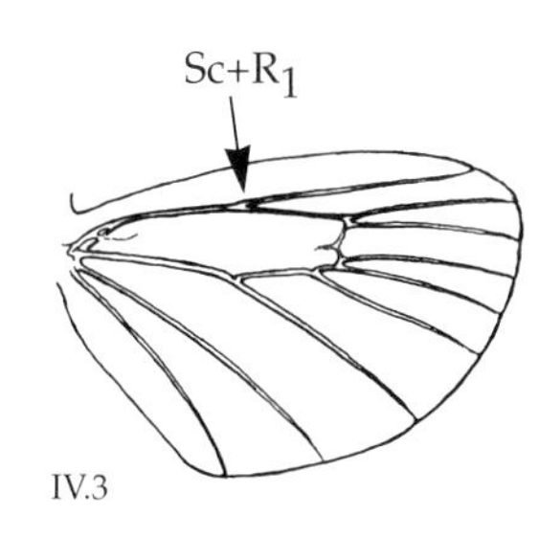

IV.3

4 Forewings emerald green, hindwings whitish; sometimes flies by day; wingspan 23–26 mm
Zygaenidae forester *Adscita statices* L.

– Wings ruby red, red/orange, white or brown; the $Sc+R_1$ vein of hindwing meets the cell about half way along its length (IV.3) tiger moths (Arctiidae) 5

5 Wings dark ruby red or red/orange, with cream and black markings. 6

– Wings predominantly grey, buff-coloured, or sooty black 7

6 Hindwings red-orange with black patches; forewings with chocolate patches on cream ground, wingspan 47–64 mm — garden tiger *Arctia caja* L.

– Fore- and hindwings dark with black dots and distinctly tinged with purplish red, wingspan 29–32 mm — ruby tiger *Phragmatobia fuliginosa* L.

7 Forewings a shade of buff to whitish with scattered grey or black spots or patches over them usually forming an oblique line from the upper outer corner; thorax and/or abdomen generally buff with dark patches on segments; wingspan 31–36 mm — buff ermine *Spilosoma lutea* Hufnagel

– Wings of females predominantly creamy-white, with black, sooty grey or brown dots; wings of males dark brown or blackish with scattered indistinct dark dots, variable; wingspan 28–31 mm — muslin moth *Diaphora mendica* Clerck

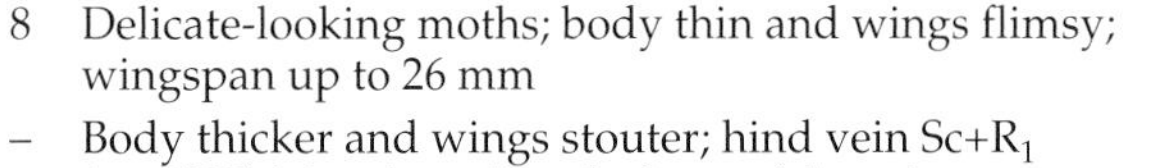

8 Delicate-looking moths; body thin and wings flimsy; wingspan up to 26 mm — 9

– Body thicker and wings stouter; hind vein $Sc+R_1$ (see IV.3) joins the cell at the base of the wing; wingspan greater than 26 mm — Noctuidae 12

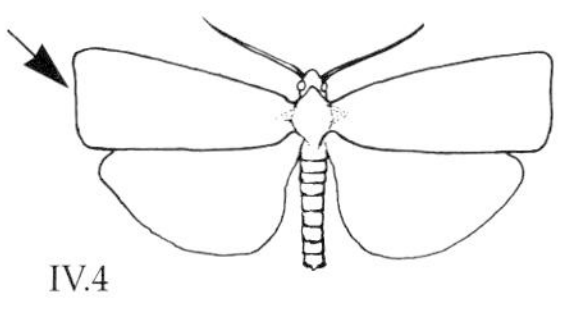

IV.4

9 Forewings almost rectangular (IV.4) — Tortricidae 10

– Forewings almost triangular (IV.5) — Geometridae 11

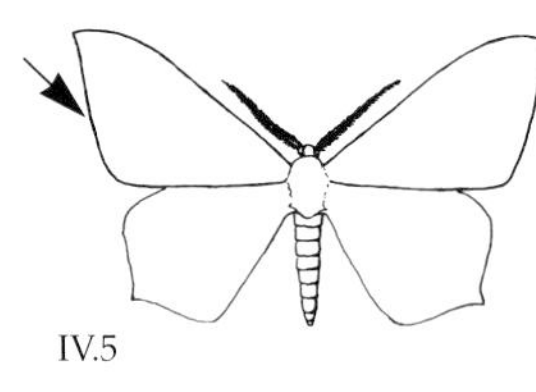

IV.5

10 Forewings whitish grey with scattered brown markings; hindwings greyish brown becoming darker towards the fringe of hairlike scales; wingspan 15–18 mm — *Cnephasia interjectana* (Haworth)

– Forewings pale ochreous or creamy white with dark brown markings — 18

11 Wings whitish edged with purplish-red; hindwings ending in points; wingspan 20–26 mm — blood-vein *Timandra griseata* Petersen

– Wings predominantly yellow or yellow-brown — 13

12 Moth active on the wing by day, forewing showing distinct white Y-shaped mark on greyish-brown ground colour; wingspan 29–41 mm — silver Y moth *Autographa gamma* (L.)

– Generally inactive by day and without Y-shaped mark on forewing — 14

13 Yellow-brown wings with fine brown markings; wingspan 22–26 mm
yellow shell *Camptogramma bilineata* L.

– Dull pale yellow with a single, often rather indistinct, black dot on all wings; wingspan 18–24 mm
plain wave *Idaea straminata* Borkhausen

14 Forewings with patches of olive green and pinkish brown, appearing bright in newly emerged specimens; hindwings barred brown; wingspan 46–49 mm
angle shades *Phlogophora meticulosa* L.

– Forewings without green and pinkish brown coloration 15

15 Forewings brown to chocolate brown with two ochreous spots; hindwings whitish; wingspan 33–39 mm
bright-line brown-eye *Lacanobia oleracea* (L.)

– Forewings not like this (may be grey) 16

16 Hindwings with dark line across middle, forewings mottled light brown with grey; wingspan 31–42 mm
rosy rustic *Hydraecia micacea* (Esper)

– Hindwings without a dark line across the middle, whitish or grey or with marginal dark stripe 17

17 Forewings grey with black patches; hindwings whitish (males) or grey (females); wingspan 33–39 mm
heart and dart *Agrotis exclamationis* (L.)

– Forewings and body grey with patches of darker speckling; hindwings white with dark edges; wingspan 34–40 mm turnip moth *Agrotis segetum* (D. & S.)

18 Forewings creamy white with dark brown markings; wingspan 14–18 mm *Olethreutes lacunana* (D. & S.)

– Forewings pale ochreous; wingspan 15–24 mm
cyclamen tortrix *Clepsis spectrana* Treitschke

Larvae

Sawfly larvae (Hymenoptera) are easily confused with Lepidoptera larvae. The sawfly species commonly found on *Rumex* is therefore included in this key. The colours and hairs on larvae can change dramatically during development. This key is designed for larvae that are fully grown when identified. Larvae that prove difficult to name may be reared through to adulthood for identification.

1 With 7 pairs of fleshy prolegs
sawfly larva (Hymenoptera: Symphyta) (pl. 3)
Ametastegia glabrata (Fallén)

– With 3 pairs of true legs on the thorax and not more than 5 pairs of fleshy prolegs on the abdomen
Lepidoptera 2

2 Feeding within a rolled leaf in a silken tube, or (in first instar) mining the leaf; very active if disturbed
Tortricidae 3

– Larva feeding within the shoot or root of the plant or openly on the plant surface 5

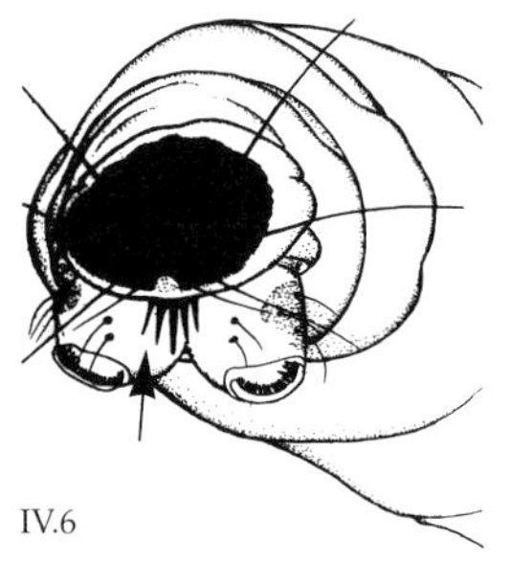

IV.6

3 First instar larva mining the leaf, later instars feeding in folded leaf; anal plate (IV.6) yellowish marked with black; anal comb with about 6 long dark prongs; abdomen varying from whitish to cream or grey-green
Cnephasia interjectana

– Larva not burrowing; anal plate differently coloured 4

This is a particularly difficult group to identify as there are many generalist species. Bradley and others (1973, 1979) should be consulted to identify adult insects reared from larvae.

4 Fully-grown larva with pale spots and stripes along the sides; anal plate cream-white mottled brown; anal comb with 6–8 prongs (pl. 3) *Clepsis spectrana*

– Larva without pale spots and side stripes, overall colour chocolate brown/black; anal plate light yellowish cream to yellow marked with brown, or all black (leaf roll pl. 4)
Olethreutes lacunana

5 Larva very or moderately hairy, with large hairs arising in tufts Arctiidae and Zygaenidae 6

– Larva not hairy or sparsely so, or covered in minute very fine hairs
Lycaenidae, Geometridae and Noctuidae 10

6 Larva covered in small tufts of hair arising from brown or pink warts; head tiny, retracted into body
forester *Adscita statices*
– Larva more hairy; head not retracted 7

7 Larva covered in long dark brown and black hairs; breathing pores (spiracles) bright white; often very active (pl. 3) garden tiger *Arctia caja*
– Larva brown or grey, covered with shorter grey or reddish-brown hairs; spiracles not bright white 8

8 Larva covered with tufts of brown hair; pale stripes along each side separating a darker back from pale brown or greyish sides (pl. 3) buff ermine *Spilosoma lutea*
– Larva covered with differently coloured hairs or without a dark back accompanied by paler sides 9

9 Body greyish brown covered with small warts from which arise yellowish brown hairs
muslin moth *Diaphora mendica*
– Larva velvety with short black/red-brown hair arising from small raised warts
ruby tiger *Phragmatobia fuliginosa*

10 Larva dorso-ventrally flattened, bright emerald green and covered in very fine, very short hairs
Lycaenidae 11
– Larva roughly cylindrical 12

11 Larva on *Rumex hydrolapathum*; bright green with retracted head covered in small white raised spots (pl. 3)
large copper *Lycaena dispar*
– Larva feeding on other *Rumex* species; smaller, with pink line along centre of back
small copper *Lycaena phlaeas*

12 Larva with 2 pairs of prolegs Geometridae 13
– Larva with 3 or 5 pairs of prolegs Noctuidae 15

13 Larva mostly brown with conspicuous V- or X-shaped markings along its back 14
– Larva with no V- or X-shaped markings but with one yellow and 2 dark lines across back
yellow shell *Camptogramma bilineata*

14 Body greyish brown with conspicuous V-shaped markings along back and swelling behind head when at rest — blood-vein *Timandra griseata*

– Body pinkish brown, wrinkled, with 4 X-shaped grey markings along back — plain wave *Idaea straminata*

15 Larva with 3 pairs of prolegs; dark stripe across side of head capsule (pl. 3) — silver Y *Autographa gamma*

– Larva with 5 pairs of prolegs — 16

16 Larva found in soil on roots or feeding inside stems near soil surface — 17

– Larva feeding on outside of the shoot — 18

17 Larva greyish pink or purple with brown spots; head yellowish-brown — rosy rustic *Hydraecia micacea*

– Larva grey and greasy in appearance without spots or purple tinge; underside yellowish-brown — turnip moth *Agrotis segetum*

18 Body reddish brown with indistinct darker stripes covered with brownish dots from which short black body hairs arise; distinct black spiracles (pl. 3) — heart and dart *Agrotis exclamationis*

– Body not like this, with either dirty whitish stripe or yellow stripe running along the side — 19

19 With scattered white dots and a yellow stripe below spiracles (pl. 3); body colour variable, from green to brown, grey or pinkish — bright-line brown-eye *Lacanobia oleracea*

– A broken white line extends along back with either side a series of dark oblique lines with whitish line running along each side (pl. 3); body colour variable from bright green to pinkish green to brown; larva fleshy — angle shades *Phlogophora meticulosa*

V Coleoptera: adult and larval beetles

This weevil key is adapted from Morris (1990), and Morris (1991) gives further information about weevil natural history and identification. Ladybirds, general predators of aphids and other small invertebrates, can be identified in Majerus & Kearns (1989) and Rotheray (1989).

Adult beetles

1 Beetle with a characteristically long snout (rostrum) which may be straight, slightly curved or folded under the head — weevils (Curculionoidea) 2

– Beetle without long rostrum — 3

2 Antennae straight, not conspicuously elbowed (V.1) — Apionidae 5

– Antennae conspicuously elbowed (V.2) — Curculionidae 14

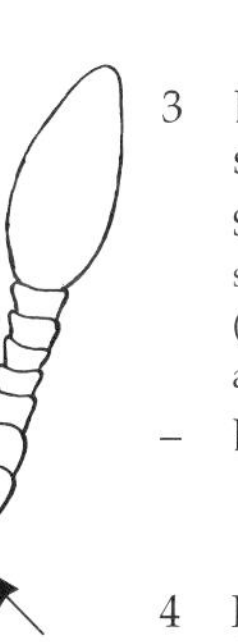

V.1

V.2

3 Beetles almost hemispherical with distinct black spots on elytra (wing cases) — ladybirds (Coccinellidae)

Several species of ladybirds may be found feeding on aphids and other small insects on docks. They can be identified using Majerus & Kearns (1989). Common species are the two spot ladybird *Adalia bipunctata* (L.) and the seven spot ladybird *Coccinella septempunctata* L.

– Beetle without distinct black spots — leaf beetles (Chrysomelidae) 4

4 Head, thorax and elytra uniformly metallic green to bronze; beetle not hairy; abdomen black (pl. 2) — *Gastrophysa viridula* Degeer

– Thorax and elytra yellowish with darker markings; beetle finely hairy (pl. 2) — *Galerucella nymphaeae* (L.)

5 Entire upper surface (except eyes and tarsal claws) red, yellowish-red or orange; legs the same colour as body; hairs sparse on both pronotum and elytra (V.4) — subgenus *Apion* 10

– At least pronotum, head and rostrum dark, black or dark brown; hairs denser — subgenus *Perapion* 6

6 Elytra blue, dark blue, greenish blue or slightly violet, bronze, or brassy, always contrasting with the black pronotum — 7

– Elytra entirely black, the same colour as pronotum and rest of body — *Apion curtirostre* Germar

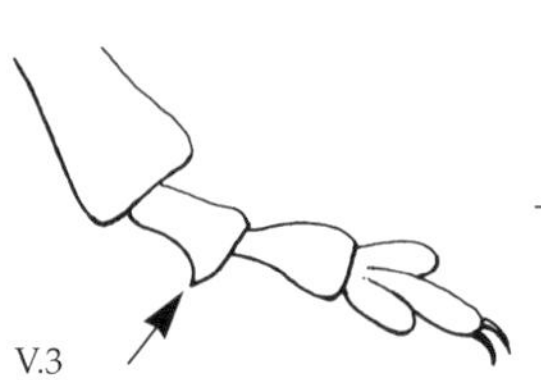

V.3

7 Elytra more elongate, 1.5–1.6 times as long as wide; scutellum (V.4) elongate; beetle larger (2.6–3.5 mm long); pronotum as long as wide or slightly longer than wide; first segment of hind tarsi of male with distinct tooth (V.3) — 8

– Elytra less elongate, 1.2–1.4 times as long as wide; scutellum about as long as wide; beetle smaller (1.6–2.4 mm long); pronotum wider than long; hind tarsi of male untoothed — 9

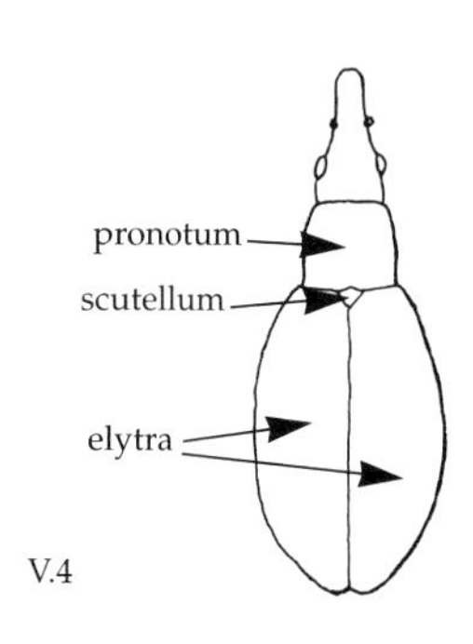

V.4

8 Rostrum shorter, straight, about as long as head, distinctly narrowing from base to tip; elytra shining (V.4) — *Apion hydrolapathi* (Marsham)

– Rostrum longer, slightly curved, longer than head, not tapering towards the tip (pl. 1) — *Apion violaceum* Kirby

9 Underside of head with large, deep pits on either side of the central hollow (also pitted); elytra wider; colour usually dark blue-black — *Apion affine* Kirby

– Underside of head without pits on either side of the central hollow; elytra narrower; colour often brighter blue-black or violet blue — *Apion marchicum* Herbst

10 Sides and 'temple' of head pitted all over; larger species, 2.7–4.4 mm long — 11

– Sides and 'temple' of head pitted immediately behind the eyes but with a wide, unpitted area at the base with lines across it; smaller species, 1.9–3.2 mm long — 12

11 Head long and narrowed in front before the eyes; length 3.6–4.4 mm (pl. 1) — *Apion frumentarium* (L.) (= *miniatum* Germar)

– Head much shorter, not narrowed in front; length 2.7–3.3 mm — *Apion cruentatum* Walton

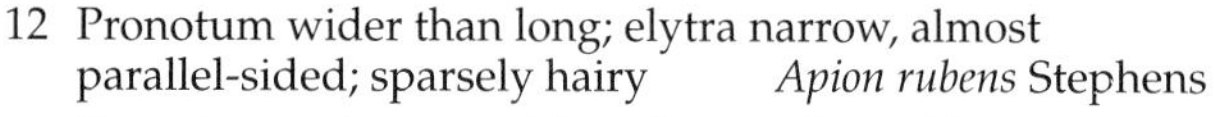

12 Pronotum wider than long; elytra narrow, almost parallel-sided; sparsely hairy — *Apion rubens* Stephens

– Pronotum as long as wide or longer than wide; elytra wider, rounded at sides; more densely hairy — 13

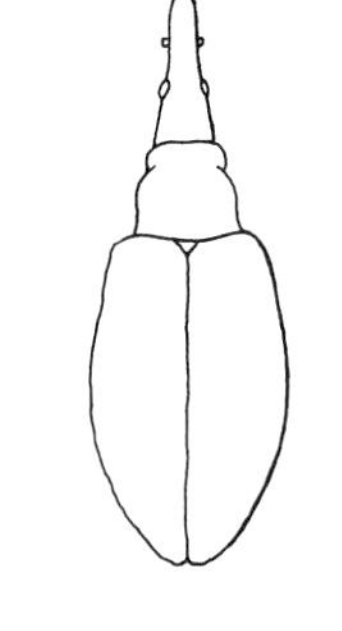

V.5

13 Rostrum almost straight, as long as pronotum; elytra broadest at middle; on *R. acetosella* on sandy areas (V.5) — *Apion rubiginosum* Grill (= *sanguineum*)

– Rostrum curved, shorter than pronotum; elytra broadest just behind the middle; on *R. acetosella* on sandy and peaty areas — *Apion haematodes* Kirby

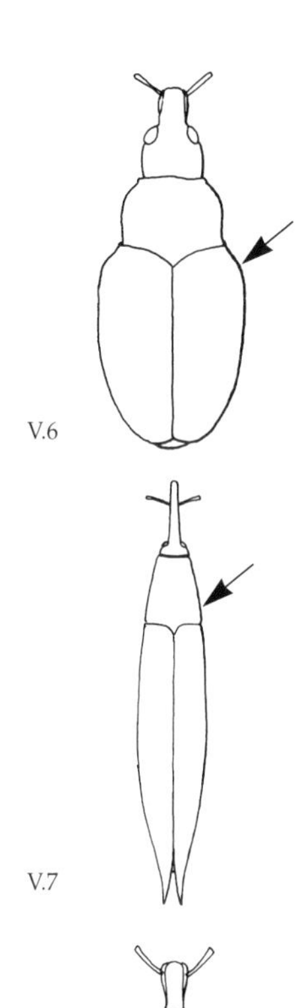

V.6

V.7

V.8

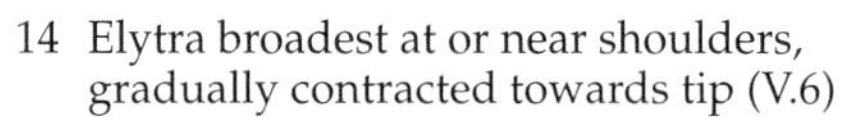

14 Elytra broadest at or near shoulders, gradually contracted towards tip (V.6) — *Rhinoncus* 15
– Elytra not shaped like this — 16

15 Smaller, 1.9–2.7 mm long; elytra distinctly longer than wide, straighter at sides, usually patterned with small patches of whitish scales; scale patch on midline of elytra usually white; elytral stripes narrower than spaces between them (pl. 1) — *Rhinoncus perpendicularis* (Reich)
– Larger, 2.5–3.7 mm long; elytra almost as wide as long, rounded at sides, usually without pattern of scales (except on elytral midline where patch is yellowish); elytral stripes almost as broad as spaces between them (pl. 1) — *Rhinoncus pericarpius* (L.)

16 Thorax broadest at the base (V.7) or, if parallel-sided, thorax and elytra covered with small hairs or scales — *Lixus* species
– Thorax narrowed at base (V.8) or, if parallel-sided for hind third, without small hairs or scales on thorax and elytra; 3.9–5.6 mm long (pl. 1) — *Hypera rumicis* L.

Beetle larvae

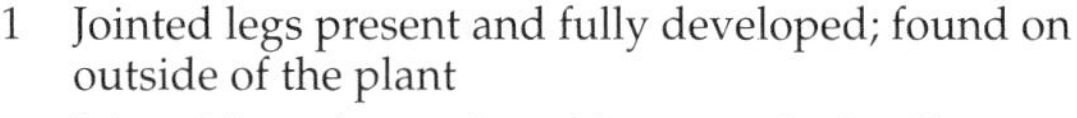

1 Jointed legs present and fully developed; found on outside of the plant — 2
– Jointed legs absent; found in stems, leaf stalks or roots (pl. 5) — weevils (Curculionoidea)

No key is provided to these larvae. The species of host plant and the insect's position within the plant will give some clues to identification (see Biology of key species, p. 17).

2 Body curved and plump, with legs shorter than width of thorax (pl. 5) — 3
– Body broader and tapering to rear, with legs as long as or longer than width of thorax (V.9) — ladybirds (Coccinellidae)

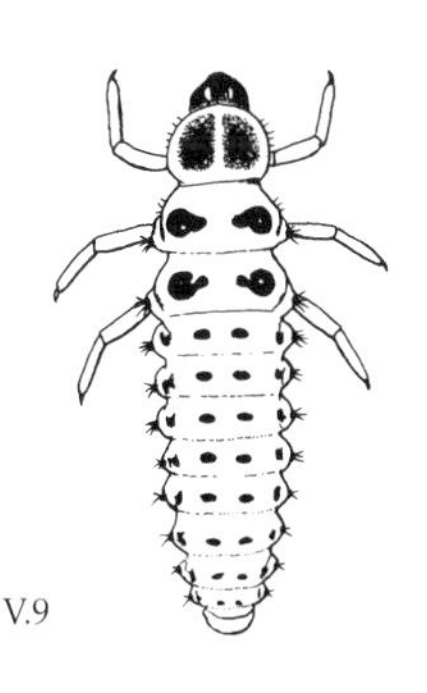

V.9

3 Larva with rows of yellow spots along back and sides (pl. 5) — *Hypera rumicis*
– Larva without rows of yellow spots along back and sides — 4

4 On *Rumex hydrolapathum* — *Galerucella nymphaeae* (pl. 5)
– Not on *R. hydrolapathum* — *Gastrophysa viridula* (pl. 5)

VI Winged adult Hymenoptera: Apocrita (ants, bees and wasps) and Symphyta (sawflies)

This key separates a sawfly found on docks and the major parasitoids of the weevils and plant-feeding flies associated with dock plants. The parasites associated with the other herbivores on dock plants have yet to be investigated, but they will have parasite complexes associated with them totalling a large number of species. Parasites should be identified from adults reared from identified host insects, as there is too little information available to identify the immature stages of parasite species.

1 Insect with a distinct waist
ants, bees and wasps (Apocrita) 2

– Insect without a very narrow waist
sawflies (Symphyta)
Ametastegia glabrata (Fallén) (pl. 3)

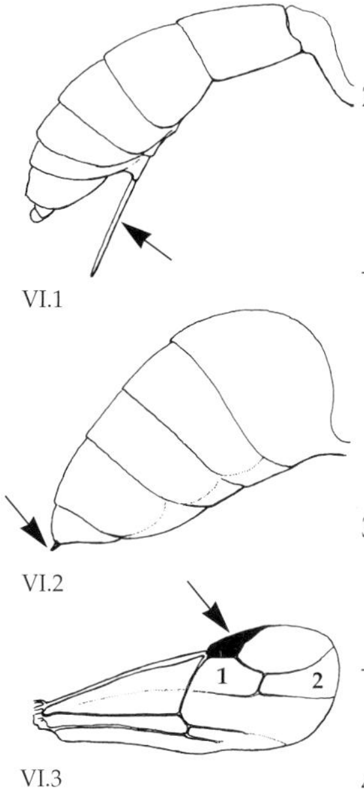

VI.1

VI.2

VI.3

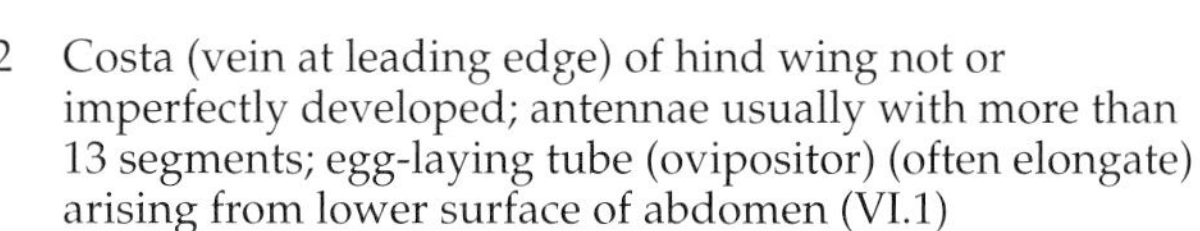

2 Costa (vein at leading edge) of hind wing not or imperfectly developed; antennae usually with more than 13 segments; egg-laying tube (ovipositor) (often elongate) arising from lower surface of abdomen (VI.1)
parasitic species 3

– Costal vein of hind wing often developed; antennae with 13 segments or fewer; ovipositor arising from tip of abdomen (VI.2) bees and ants

Ants may be found on docks either searching for prey or tending colonies of aphids, typically *Aphis rumicis*. They may be identified using Skinner & Allen (1996).

3 Forewing with distinct black spot at foremargin (pterostigma,VI.3); antennae usually with more than 16 segments
superfamily Ichneumonoidea: Braconidae 4

– Forewing with no true pterostigma; antennae with not more than 16 segments 6

4 Mouth opening present between distinctly notched clypeus (lowest part of face) and the mandibles; fore tibia without spines on front surface; forewing with 3 submarginal cells (two are labelled on VI.3)
Colastes braconius Haliday

– Mouth opening not present between clypeus and mandibles 5

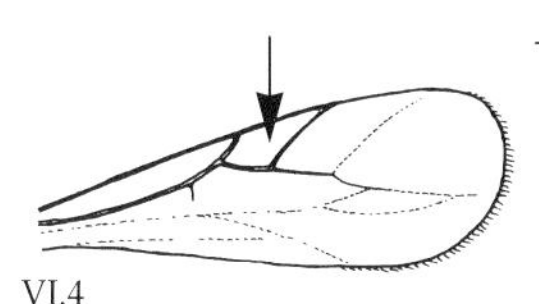
VI.4

5 Mandibles not meeting even when closed, always with 3 or more teeth *Dapsilarthra* (*Adelurola*) *florimela* Haliday

– Mandibles meeting when closed, always with 2 teeth
Opiinae
Opius rufipes Wesmael
Biosteres carbonarius (Nees)
B. impressus Wesmael

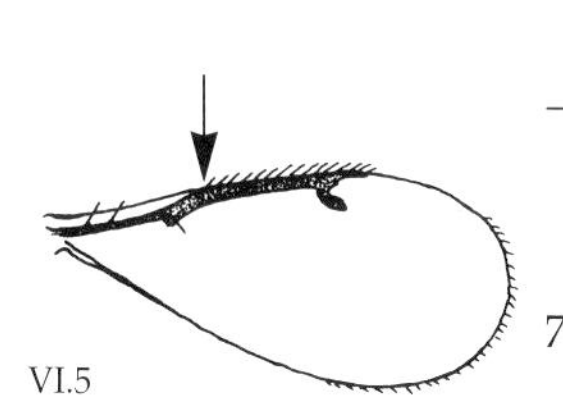
VI.5

6 Pronotum (front part of thorax viewed from above; see VI.6) pointed at the sides below; forewings with cell R1 more or less complete (VI.4); pronounced raised cup on disc of scutellum superfamily Cynipoidea, family Eucoilidae *Trybliographa gracilicornis* (Cameron)

– Pronotum rounded at the sides below; cell R1 not defined by proper veins or absent (VI.5)
superfamily Chalcidoidea 7

7 Axillae (hard plates at base of wing) not or little advanced in front of front margin of scutellum 8

– Axillae advanced strongly in front of front margin of scutellum (VI.6) Eulophidae 9

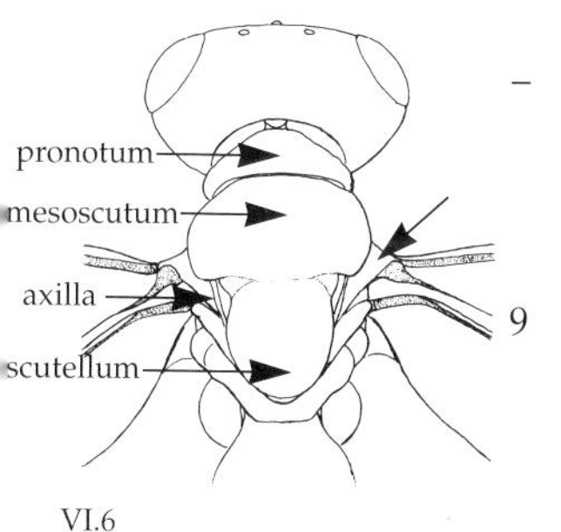

VI.6

8 Pronotum short, very narrow at front; colours usually metallic Pteromalidae *Chlorocytus laogore* (Walker)
Lamprotatus splendens Westwood
Seladerma breve Walker
Skeloceras truncatum (Fonscolombe)

– Pronotum about as long as wide, scarcely narrower at front than mesoscutum (part of thorax in front of scutellum); colours black or yellowish
Eurytomidae *Eurytoma curculionum* Mayr

9 Scutellum bears two hairs; only 2 hairs on upper surface of forewing submarginal vein
Chrysocharis nephereus (Walker)
Entedon rumicis Graham (pl. 1)
E. hercyna Walker

– Scutellum bears four hairs; more than two hairs on upper surface of forewing submarginal vein *Cirrospilus* species

VII Diptera: adult and larval flies

This key is designed to separate the main plant-feeding species and common predators of aphids on dock plants. There are about 100 species of hoverflies which consume aphids and one which can often be found feeding on first-instar *Gastrophysa* larvae. The following key includes some of the species commonly encountered on low vegetation, but adults should be reared from larvae or pupae collected on the plants and identified in Stubbs & Falk (1983), Gilbert (1986) or Rotheray (1989). Larval hoverflies are covered by Rotheray (1993). Diptera often show sexual dimorphism. Males and females may differ in patterning and in the size and alignment of the eyes, particularly in predatory and parasitic species.

VII.1

Adult flies

1 Delicate fly with antennal segments looking like beads in a necklace, ringed with hairs (VII.1)
Nematocera: Cecidomyiidae
Contarinia acetosellae (Rübsaamen) (pl. 1)

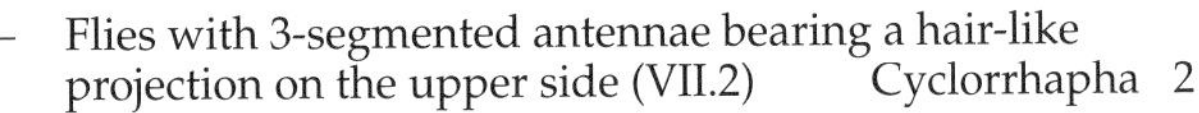

– Flies with 3-segmented antennae bearing a hair-like projection on the upper side (VII.2) Cyclorrhapha 2

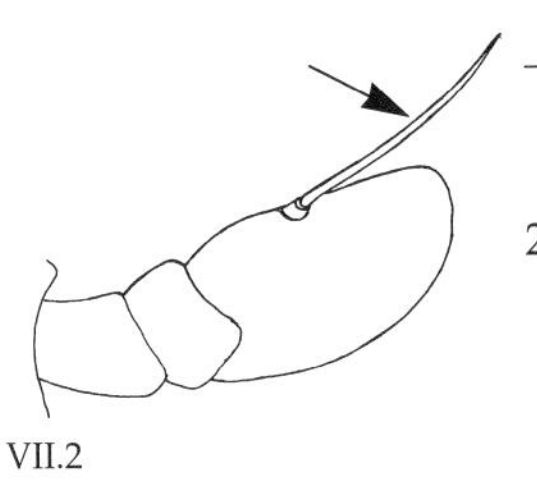

VII.2

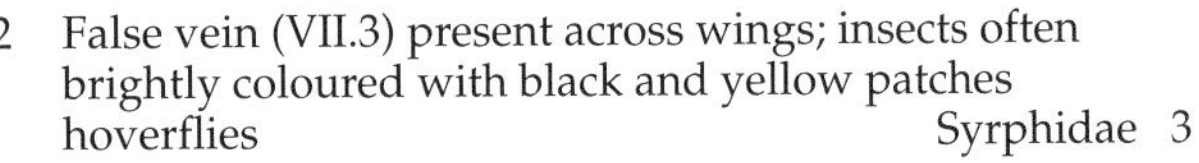

2 False vein (VII.3) present across wings; insects often brightly coloured with black and yellow patches hoverflies Syrphidae 3

Among the species commonly found on low vegetation are:

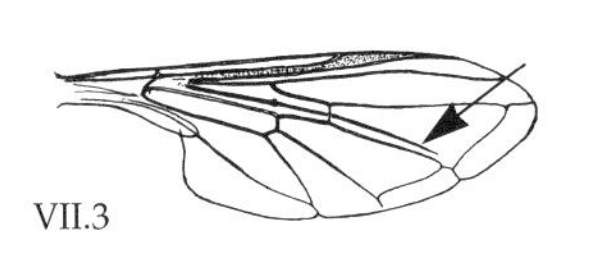

VII.3

Platycheirus scutatus (Meigen) (pl. 2) (Face region black; abdomen wide; male with tibia tufted with black hairs; female with yellowish mark on underside of last antennal segment; dorsal plates with yellow markings, hind margins of yellow marks on second dorsal plate straight not oblique.)

Sphaerophoria scripta (L.) (Distinct yellow stripe on top of thorax, male abdomen longer than wing.)

Metasyrphus luniger (Meigen) (pl. 2) (Line of black hairs on side margin of dorsal plates on abdomen.)

– No false vein; insect not brightly coloured 3

3 Frontalia with crossed bristles (VII.4) Anthomyiidae 4

– Frontalia without crossed bristles; frons equally broad in males and females Scathophagidae (= Cordyluridae)
Norellisoma spinimanum (Fallén) (pl. 2)

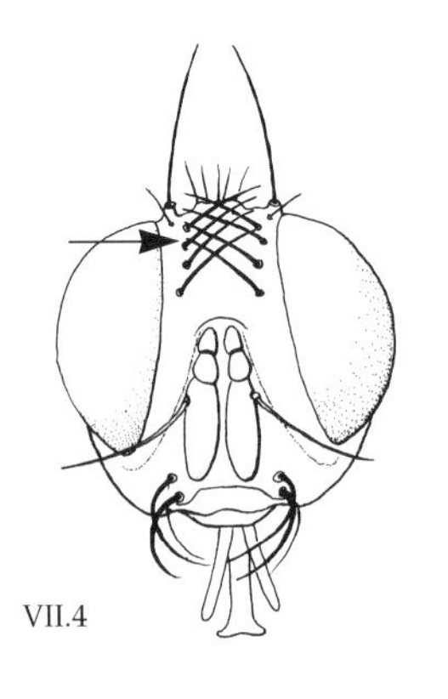

VII.4

4 Fly 5–6 mm long; thorax grey; abdomen reddish yellow
Pegomya nigritarsis (Zetterstedt) (pl. 2)

– Fly 6–8 mm long; thorax grey; abdomen reddish brown; palps yellow with black ends
Pegomya bicolor (Hoffmannsegg) (pl. 2)

Fly larvae

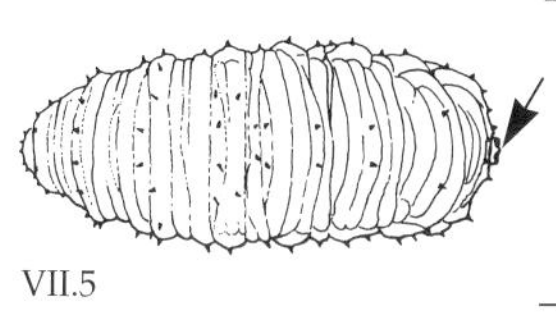
VII.5

1 Larva found living freely on plant surface; larvae slug-like, flattened, grey, green or brown with warty tubercles; preying on aphids or *Gastrophysa* eggs; hind end of body with paired spiracles arising from single tube-like fleshy projection (VII.5) hoverfly larvae (Syrphidae) (pl. 5)
(for further identification see Rotheray 1993)

– Found inside plant (flower head, leaf stalk, leaf or stem); without paired spiracles arising from a single projection 2

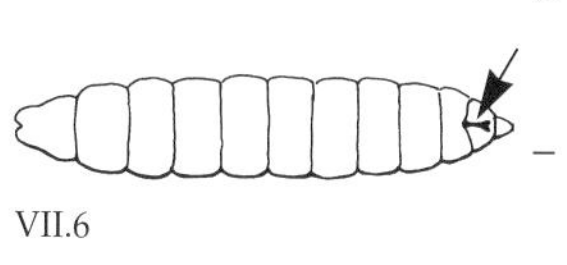
VII.6

2 Larva in blotch mines in leaves (pl. 4) or boring large leaf veins, stems or leaf stalks, with frass often visible at the exit (pl. 4); larva whitish 3

– Larva in gall in inflorescence, white or pinkish; final-instar larva has spatula visible below (VII.6) (use a microscope) Cecidomyiidae *Contarinia acetosellae*

VII.7

3 Larva in leaf stalk or large leaf veins, often with dark frass accumulating at entrance hole, and sometimes causing distortion of leaf or stem; anal lobes separate and unbranched Scathophagidae (= Cordyluridae)
Norellisoma spinimanum (pl. 5)

– Larva in blotch mine in leaf; anal lobes 2-branched (VII.7)
Anthomyiidae *Pegomya* species (pl. 5)

5 Techniques

Collecting dock-feeding insects

In order to identify many insects it is necessary to kill them so that diagnostic details of their structure can be examined. They can then be kept, labelled with locality, date and Latin name, in a reference collection against which other species can be compared. It is important to keep accurate detailed records of where and when specimens were collected and the host they were feeding on. This enormously increases the scientific value of any reference collection.

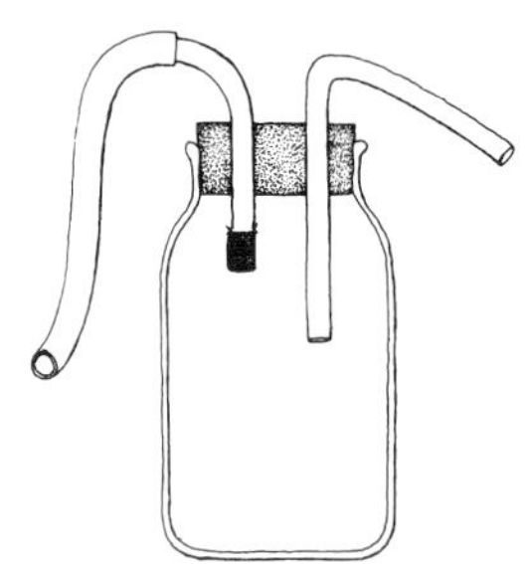

Fig. 25. A pooter.

Nets and plastic boxes and tubes are invaluable for catching the insects. Small insects can be collected from plant surfaces by means of a pooter (fig. 25). This consists of a small glass vessel fitted with a rubber bung through which pass two tubes. Insects are sucked up into one tube when the operator sucks through the other (which is covered by gauze at the other end to prevent insects from being sucked into the mouth).

Hard-bodied adult insects can be killed quickly in a vessel containing chopped cherry laurel (*Prunus laurocerasus* L.) leaves, which give off a weak cyanide vapour. The specimens can be set or pinned in a suitable position so that important features for identification and comparison with other specimens are exposed. Instructions for this procedure and further details concerning collecting and display are provided in many of the *Naturalists' Handbooks* and in Chinery (1986).

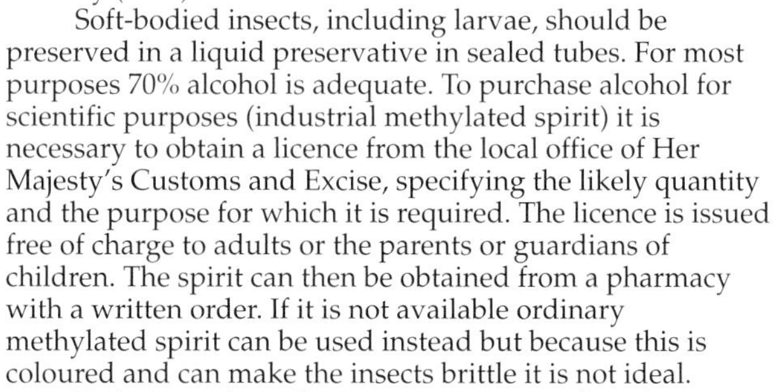

Soft-bodied insects, including larvae, should be preserved in a liquid preservative in sealed tubes. For most purposes 70% alcohol is adequate. To purchase alcohol for scientific purposes (industrial methylated spirit) it is necessary to obtain a licence from the local office of Her Majesty's Customs and Excise, specifying the likely quantity and the purpose for which it is required. The licence is issued free of charge to adults or the parents or guardians of children. The spirit can then be obtained from a pharmacy with a written order. If it is not available ordinary methylated spirit can be used instead but because this is coloured and can make the insects brittle it is not ideal.

Techniques for collecting, preserving and studying aphids as well as ecological information on this important group are given in Blackman (1974).

Equipment for collecting, preserving and displaying insects can be obtained from entomological suppliers (addresses p. 47).

Growing plants and rearing insects

It is quite easy to grow *Rumex* plants in pots in a greenhouse where they will grow well in horticultural compost. Wild-collected plants or even root fragments can be grown on in plots outside or on benches in the laboratory. Seeds can be bought from suppliers (addresses p. 47) or collected from the field although wild-collected seed sometimes results in infertile hybrid plants.

Gastrophysa beetles can be cultured on living plants inside cages made of acrylic sheet with a gauze top (fig. 26). As well as permitting close observation of the beetle in all its stages, these cultures may be used to complement field studies and to carry out research projects. Gravid females (those with the abdomen distended with eggs) can be placed in a plastic sandwich box containing detached, healthy *Rumex* leaves on a base of damp filter paper or sand, where they will lay eggs on the leaves. As soon as the eggs hatch, transfer the first-instar larvae to fresh leaves. Replenish the leaves every few days. The beetles will undergo two moults and then pupate in the box. The next generation of adults will shortly emerge. The whole life cycle will take about 4 weeks at 15°C and less time at higher temperatures. In this way, it is possible to keep beetles alive through most of the year. Some Lepidoptera can be readily reared and bred in captivity and are available either from commercial suppliers or through the Entomological Livestock Group. Books giving detailed practical information on breeding butterflies and moths include Friedrich (1986) and Stone & Midwinter (1975).

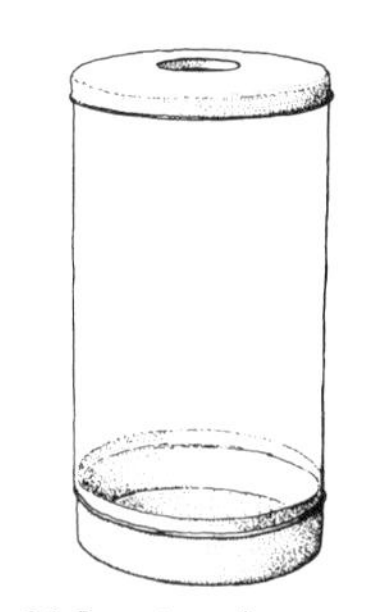

Fig. 26. Insect rearing cage.

Investigations

We hope this introduction will encourage readers to embark on field and laboratory investigations of their own to add to our knowledge of the insect community associated with dock plants. Observations made in the field are most productive if linked to specific questions or hypotheses. Trained ecologists will always try to quantify the information which they collect in such a way that the results can be used to test specific hypotheses and suggest further observations or experiments. Thus, for example, instead of simply recording the fact that *Philaenus spumarius* has a number of colour forms, the investigator will record the frequencies (relative numbers) of each morph found in different habitats or on different host plants and analyse this information to see if there are recognisable patterns. In short, if the natural historian is interested in variety and diversity, the ecologist is interested in patterns and processes. A good field worker employs a mixture of these approaches. Although more is known about the natural history of Britain than anywhere else in the world, much remains to be discovered. The distributions of many species of insects other than Lepidoptera are poorly known, and knowledge of the biology of the animals is incomplete. As well as scope for identification of overlooked insects associated with docks there is scope for detailed information to be collected on distribution, biology and on how the insects interact with the host plants. Where measurements of any sort are taken their thorough comparison generally requires some use of statistics from simply calculating means to more complex calculations. Chalmers & Parker (1989) is a good source of information on using statistics in ecological projects. Advice on how to write and publish scientific papers can be found in Day (1988).

Several journals publish results of small scale experiments and field observations. These include the *Bulletin of the Amateur Entomologists' Society*, published by the Amateur Entomologists' Society and the *Journal of the British Entomology and Natural History Society*, published by The British Entomological and Natural History Society. These societies publish useful handbooks and hold exhibitions and meetings. The *Entomologist's Monthly Magazine* also publishes short reports of findings. A copy of the journal should be consulted for the publishing conventions, article length and style before the manuscript is prepared.

Some useful addresses

Several commercial seed merchants stock seed of a range of dock species, although seed with high viability can be collected from wild plants. The Entomological Livestock Group produces a newsletter advertising live insect material, mainly Lepidoptera, which can be used in investigations. The British Ecological Society dispenses small project grants for ecological research of publishable quality, for example allowing travel to field sites. Details of membership of the society and of small project grants can be obtained from the Society's office.

Local natural history societies often provide a good opportunity to learn about other areas of ecology and members generally have expertise in identifying plant and insect material. Local libraries often keep addresses of the natural history societies in the locality. In many cases the curators of natural history departments in local museums will help to identify specimens and have named reference collections with which insects can be compared. Some keep records of the regional distribution of particular insect groups.

Entomological societies and interest groups

Amateur Entomologists' Society, P.O. Box 8774,
London SW7 5ZG

British Ecological Society, 26 Blades Court, Deodar Road,
Putney, London SW15 2NU
E-mail:general@ecology.demon.co.uk
http://www.demon.co.uk/bes

British Entomological and Natural History Society, Pelham-Clinton Building, Dinton Pastures Country Park, Davis Street, Hurst, Reading, Berks RG10 0TH

Entomological Livestock Group, 11 Rock Gardens,
Aldershot, Hants GU11 3AD

Royal Entomological Society, 41 Queen's Gate,
London SW7 5HR

Suppliers of entomological and natural history equipment

Watkins & Doncaster, The Naturalists, P.O. Box 5, Cranbrook,
Kent TN18 5EZ

Natural history booksellers

E. W. Classey Ltd, Oxford House, Marlborough Street, Faringdon, Oxon SN7 7JP
(new and secondhand)

Little Holcombe Books, 10 Lumb Carr Avenue, Ramsbottom, Bury BL0 9QG
(secondhand)

Natural History Book Service, 2, Wills Rd, Totnes, Devon TQ9 5XN

The Richmond Publishing Co. Ltd, PO Box 963, Slough SL2 3RS (for Naturalists' Handbooks and AIDGAP keys)

Suppliers of seeds

Emorsgate Seeds, The Pea Mill, Market Lane, Terrington St Clement, Norfolk PE34 4HR

Herbiseed, The Nurseries, Billingbear Park, Wokingham RG40 5RY

References and further reading

Finding books

Local reference libraries may have good collections of books on the identification of insects, but more specialised scientific journals are generally only held by university libraries and those of scientific societies. However they should be available from the lending division of the British Library through local libraries. University libraries also often allow local people to consult their books and journals and should be approached. Detailed keys to British insects are still being written and updated but most common species have now been described in *Handbooks for the Identification of British Insects* although there is still much to discover about their biology. The keys are available from The Royal Entomological Society. Naturalists' Handbooks and AIDGAP keys (indicated by an asterisk) are available from The Richmond Publishing Co. Ltd, P.O. Box 963, Slough SL2 3RS.

Barbattini, R., Zandigiacomo, P. & Parmegiani, P. (1986). Indagine preliminare sui fitofagi di *Rumex obtusifolius* L. e *Rumex crispus* L. in vigneti del Fruili. [In Italian, with English summary.] *Redia* **69**, 131–142.

Bentley, S. & Whittaker, J.B. (1979). Effects of grazing by a chrysomelid beetle *Gastrophysa viridula* on competition between *Rumex obtusifolius* and *Rumex crispus*. *Journal of Ecology* **67**, 79–90.

Blackman, R. (1974). *Invertebrate Types: Aphids*. London: Ginn & Co. Ltd.

Blower, J.G. (1958). *British millipedes (Diplopoda) with keys to the species*. The Linnean Society of London, Synopses of the British Fauna no. 11.

Bradley, J.D., Tremewan, W.G. & Smith, B.A. (1973, 1979). *British Tortricoid Moths*. London: British Museum (Natural History)/Ray Society 147, 153.

Brooks, G.L. & Whittaker, J.B. (in press). Responses to elevated CO_2 of multiple generations of *Gastrophysa viridula* feeding on *Rumex obtusifolius*. *Global Change Biology*.

Carter, D.J. & Hargreaves, B.A. (1986). *Field Guide to Caterpillars of Butterflies and Moths in Britain and Europe*. London: Collins.

Chalmers, N. & Parker, P. (1989). *Fieldwork and Statistics for Ecological Projects*. (2nd edn.) Shrewsbury: Field Studies Council.

Chamberlin, T.R. (1933). Some observations on the life history and parasites of *Hypera rumicis* (L.) (Coleoptera: Curculionidae). *Proceedings of the Entomological Society of Washington* **35**, 101–109.

Chancellor, R.J. (1959). *Identification of Seedlings of Common Weeds*. London: HMSO Bulletin No. 179.

Chinery, M. (1986). *A Field Guide to the Insects of Britain and Northern Europe*. London: Collins. (2nd edn.)

Chinery, M. (1993). *Insects of Britain and Northern Europe*. (3rd edn). London: Collins

Cideciyan, M.A. & Malloch, A.J.C. (1982). Effects of seed size on the germination, growth and competitive ability of *Rumex crispus* and *Rumex obtusifolius*. *Journal of Ecology* **70**, 227–232.

Cottam, D., Whittaker, J.B. & Malloch, A.J.C. (1986). The effects of chrysomelid beetle grazing and plant competition on the growth of *Rumex obtusifolius*. *Oecologia* **70**, 452–456.

Day, R.A. (1988). *How to Write and Publish a Scientific Paper*. (3rd edn.) Phoenix: Oryx Press

De Gregorio, R.E., Ashley, R.A., Adams, R.G., Streams, F.A. & Schaefer, C.W. (1991). Biocontrol potential of *Hypera rumicis* (L.) (Coleoptera: Curculionidae) on curly dock (*Rumex crispus* L.). *Journal of Sustainable Agriculture* **2**, 7–24.

Disney, R.H.L. (1976). The pre-adult stages of *Norellisoma spinimanum* (Fallén) (Dipt., Cordyluridae) and a parasitoid (Hym., Pteromalidae) of the same. *Entomologist's Gazette* **27**, 263–267.

Docters van Leeuwen, W.M. (1982). *Gallenboek.* revised and edited by A.A. Wiebes-Rijks & G. Houtman. Zutphen: B.V.W.J. Thieme & Cie. [In Dutch.]

Eason, E.H. (1964). *Centipedes of the British Isles.* London: Frederick Warne.

Evans, G.O. (1992). *Principles of Acarology.* Wallingford: CAB International.

Finch, S. & Jones, T.H. (1989). An analysis of the deterrent effect of aphids on cabbage root fly (*Delia radicum*) egg-laying. *Ecological Entomology* **14**, 387–391.

Fjellberg, A. (1980) *Identification keys to Norwegian Collembola.* Aas: Norsk Entomolgisk Forening.

Foster, L. (1989). The biology and non-chemical control of dock species *Rumex obtusifolius* and *R. crispus. Biological Agriculture and Horticulture* **6,** 11–25.

Freese, G. (1995). Structural refuges in two stem-boring weevils on *Rumex crispus. Ecological Entomology* **20**, 351–358.

Friedrich, E. (1986). *Breeding Butterflies and Moths, A Practical Handbook for British and European Species.* Colchester: Harley Books.

*Gilbert, F.S. (1986). *Hoverflies.* Naturalists' Handbooks 5. Slough: The Richmond Publishing Co. Ltd.

Godfray, H.C.J. (1986). Clutch size in a leaf mining fly *Pegomya nigritarsis* (Diptera: Anthomyiidae). *Ecological Entomology* **11**, 75–81.

Harper, G.A. (1974). The classification of adult colour forms of *Philaenus spumarius* (L.) (Homoptera: Insecta). *Zoological Journal of the Linnean Society* **55**, 177–192.

Hatcher, P.E., Paul, N.D., Ayres, P.G. & Whittaker, J.B. (1994). The effect of a foliar disease (rust) on the development of *Gastrophysa viridula* (Coleoptera: Chrysomelidae). *Ecological Entomology* **19**, 349–360.

Hillyard, P.D. & Sankey, J.H.P. (1989). *British harvestmen.* The Linnean Society of London, Synopses of the British Fauna no. 4.

*Hopkin, S. (1991). A key to the woodlice of Britain and Ireland. *Field Studies* **7**, 599–650 (An AIDGAP key).

Hopkins, M.J.G. (1984). The parasite complex associated with stem-boring *Apion* (Col., Curculionidae) feeding on *Rumex* species (Polygonaceae). *Entomologist's Monthly Magazine* **120**, 187–192.

Hopkins, M.J.G. & Whittaker, J. B. (1980a). Interactions between *Apion* species (Coleoptera: Curculionidae) and Polygonaceae. I. *Apion curtirostre* Germ. and *Rumex acetosa* L. *Ecological Entomology* **5**, 227–239.

Hopkins, M.J.G. & Whittaker, J.B. (1980b). Interactions between *Apion* species (Coleoptera: Curculionidae) and Polygonaceae. II. *Apion violaceum* Kirby and *Rumex obtusifolius* L. *Ecological Entomology* **5**, 241–274.

Kent, D.H. (1992). *List of Vascular Plants of the British Isles.* London: Botanical Society of the British Isles.

Kloet, G.S. & Hincks, W.D. (1964–1978). *A Check List of British Insects.* Part 1: Small orders and Hermiptera (1964); Part 2: Lepidoptera (1972); Part 5: Diptera and Siphonaptera (1975); Part 3: Coleoptera (1977); Part 4: Hymenoptera (1978). *Handbooks for the Identification of British Insects* **11**. London: Royal Entomological Society of London.

Lousley, J.E. & Kent, D.H. (1981). *Docks and Knotweeds of the British Isles.* B.S.B.I. Handbook 3. London: Botanical Society of the British Isles.

Mabey, R. (1996). *Flora Britannica*. London: Sinclair-Stevenson.

Macleod, R.D. (1952). *Key to the names of British plants*. London, Sir Isaac Pitman & Sons Ltd.

*Majerus, M. & Kearns, P. (1989). *Ladybirds*. Naturalists' Handbooks 10. Slough: The Richmond Publishing Co. Ltd.

Morris, M.G. (1990). Orthocerous Weevils. Coleoptera: Curculionoidea (Nemonychidae, Anthribidae, Urodontidae, Attelabidae and Apionidae). *Handbooks for the Identification of British Insects* **5** (16), 1–108. London: Royal Entomological Society.

*Morris, M.G. (1991). *Weevils*. Naturalists' Handbooks 19. Slough: The Richmond Publishing Co. Ltd.

Paviour-Smith, K. & Whittaker, J.B. (1967). A key to the Major British Groups of Terrestrial Invertebrates. In *The Teaching of Ecology* (ed. J. Lambert), Symposium of the British Ecological Society 7, pp 24–32. Oxford: Blackwell Scientific Publications.

*Plant, C. (1997). A key to the adults of British lacewings and their allies (Neuroptera, Megaloptera, Rhaphidioptera and Mecoptera). *Field Studies* **9**, 179–270 (an AIDGAP key).

Roberts, M.J. (1995). *Spiders of Britain and Northern Europe*. London: HarperCollins.

*Rotheray, G.E. (1989). *Aphid Predators*. Naturalists' Handbooks 11. Slough: The Richmond Publishing Co. Ltd.

Rotheray, G.E. (1993). Colour guide to hoverfly larvae (Diptera, Syrphidae). *Dipterists Digest* **9**, 1-156.

Salisbury, E. (1964). *Weeds and Aliens*. London: Collins, The New Naturalist, no. 43.

Salt, D.T. & Whittaker, J.B. (1995). Elevated carbon dioxide affects leaf-miner performance and plant growth in docks (*Rumex* spp.). *Global Change Biology* **1**, 153–156.

*Skinner, G. & Allen, G.W. (1996). *Ants*. Naturalists' Handbooks 24. Slough: The Richmond Publishing Co. Ltd.

Southwood, T.R.E. & Leston, D. (1959). *Land and Water Bugs of the British Isles*. London: Frederick Warne.

Stace, C. (1991). *New Flora of the British Isles*. Cambridge: Cambridge University Press.

Stone, J.L.S. & Midwinter, H.J. (1975). *Butterfly Culture, A Guide to Breeding Butterflies, Moths and Other Insects*. Poole: Blandford Press.

Stroyan, H.L.G. (1984). Aphids-Pterocommatinae and Aphidinae (Aphidini) Homoptera, Aphididae. *Handbooks for the Identification of British Insects* **2** (6), 1–232. London: Royal Entomological Society.

Stubbs, A.E. & Falk, S.J. (1983). *British Hoverflies. An Illustrated Identification Guide*. London: British Entomological and Natural History Society.

Thomas, J. & Lewington, R. (1991). *The Butterflies of Britain and Ireland*. London: Dorling Kindersley.

Whittaker, J.B. (1969). The biology of Pipunculidae (Diptera) parasitising some British Cercopidae (Homoptera). *Proceedings of the Royal Entomological Society of London (A)* **44**, 17–24.

Whittaker, J.B. (1982). The effect of grazing by a chrysomelid beetle *Gastrophysa viridula* on growth and survival of *Rumex crispus* on a shingle bank. *Journal of Ecology* **70**, 291–296.

Whittaker, J.B. (1992). Green plants and plant-feeding insects. *Journal of Biological Education* **26**, 257-262.

Whittaker, J.B. (1994). Physiological responses of leaves of *Rumex obtusifolius* to damage by a leaf miner. *Functional Ecology* **8**, 627–630.

Whittaker J.B., Kristiansen, L.W., Mikkelson, T.N. & Moore, R. (1989). Responses to ozone of insects feeding on a crop and a weed species. *Environmental Pollution* **62**, 89–101.

Index